ELECTRONIC PROJECTS
USING
SOLAR CELLS

D1329247

ELECTRONIC PROJECTS
USING
SOLAR CELLS

by
OWEN BISHOP

BERNARD BABANI (publishing) LTD
THE GRAMPIANS
SHEPHERDS BUSH ROAD
LONDON W6 7NF
ENGLAND

Although every care is taken with the preparation of this book, the publishers or author will not be responsible in any way for any errors that might occur.

© 1981 BERNARD BABANI (publishing) LTD

First Published — October 1981

British Library Cataloguing in Publication Data
Bishop, Owen
   Electronic projects using solar cells.
   1. Solar batteries
   2. Electronic apparatus and appliances — Amateurs'
     manuals
   I. Title
   621.31'244   TK290

   ISBN 0 85934 057 0

Printed and Bound in Great Britain by Cox and Wyman Ltd., Reading

PREFACE

In designing the projects for this book it has been assumed that the average constructor will not wish to invest time and money in building large arrays of solar cells. After all, given a sufficiently large array, almost *any* circuit can be powered from sunlight so there would be no particular need to write a book on such a subject. All but one of the circuits in the book are specially designed to operate at low voltage and to consume small currents. This is what suits them to using solar power from a small array of cells. Many can operate on only two or three of the smallest kind of cell. As a consequence of this policy two further points arise. One is that the circuits tend to be built around the least possible number of components, making them simple for the beginner to construct. The other point is that the circuits are economical of power from *any* source, so that, if the reader prefers not to use solar cells, they can be powered by small batteries, button-cells and the like, just as easily. In short, this is a book of simple, low-cost, low-power circuits which have applications in and around the home. Whether or not they are powered by the sun, by dry cells or any other means is a matter for the reader to decide.

Owen Bishop

CONTENTS

CHAPTER 1

How Solar Cells Work

In this book we deal with the silicon solar cell, the type most commonly in use at present. The silicon solar cell belongs to the class of device known as a *photovoltaic cell*. This class also includes photodiodes and phototransistors. When light energy falls on a photovoltaic cell, an emf is generated between its terminals. The cell *generates* electric power, in contrast to a photo-conductive cell (for example the cadmium sulphide light dependent resistor) which merely shows a change in electrical resistance. Thus solar cells can be used as a source of electrical power, a fact emphasised by the electrical symbol for the device (Fig. 1).

The cell consists of a wafer of p-type silicon, on the surface of which a thin n-type layer is created by diffusion (Fig. 2). This forms a p-n junction, just as in an ordinary semiconductor diode, and a depletion region forms on either side of the junction. This is the result of excess electrons from the n-type silicon crossing over to the p-type silicon and filling the excess holes there. This leaves ionised atoms on either side of the junction. They are immobile, for they are fixed in their positions in the crystal lattice. The effect is to produce an

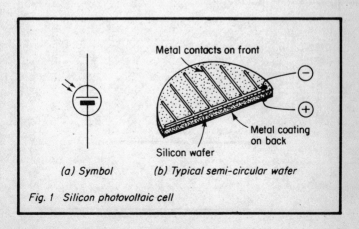

(a) Symbol   *(b) Typical semi-circular wafer*

Fig. 1 Silicon photovoltaic cell

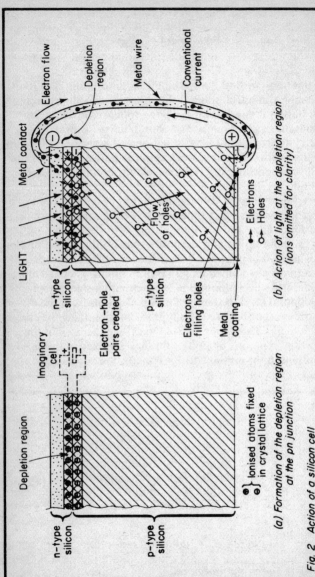

Fig. 2 Action of a silicon cell

(a) Formation of the depletion region at the pn junction

(b) Action of light at the depletion region (ions omitted for clarity)

2

electric field across the depletion region, having a potential difference of about 0.6v. If light falls on the depletion region, which is necessarily at the surface of the material, electron-hole pairs are created. The electrons and holes migrate under the influence of the electric field. If the terminals of the cell are connected by wire or the external circuit is completed in some other way, a current flows. What actually happens is that a flow of electrons in the wire (or other connecting circuit) removes the excess electrons from the n-type material and supplies electrons to fill the excess holes in the p-type material. If we think of this in terms of conventional current, we have a flow of current from positive (the back surface of the slice) to negative (the illuminated front surface of the slice).

A similar effect is found in the silicon photodiode, but in this the current is small because of the small cross-sectional area (and thus high resistance) of the device. In the photo-transistor the creation of electron-hole pairs in the base layer produces what is in effect a small base current. This causes an increased collector current to flow. In the silicon cell the area across which the current flows is much larger, a matter of several square centimetres. In this way the device is con-structed with relatively low resistance, especially for current production. Despite this, solar cells do not have high efficiency in the conversion of light energy to electrical energy. On an average sunny day in the latitude of Britain the power of incident sunlight is $500W/m^2$, sometimes more. A small wafer with area $10cm^2$ would receive 0.5 watts or 500 milliwatts. A cell of this size normally generates electricity at about 50 milliwatts which is only 10% efficiency. Part of the loss of power is due to the electrical resistance of the cells, electrical energy being converted to thermal energy in the cell. Part of the incident light energy is reflected from the wafer before it reaches the depletion region, so is never made available for generating electricity. In addition, a proportion of the electron-hole pairs recombine in the depletion region before they have been swept into the regions on either side. This too represents a waste of the incident light energy.

Researchers are busily engaged in trying to attain higher efficiencies of energy conversion. New materials and new methods of manufacture and construction are being tried and the prospects of success are high.

CHAPTER 2

Solar Cells in Practice

As explained in Chapter 1, the maximum emf of a single solar cell is approximately 0.6v. Under normal daylight conditions a lower emf (0.45v – 0.5v) is more usual. The amount of current that can be supplied depends mainly on the area of the cell. Commonly available cells provide between 20mA and 500mA. In general, the cost of the cell is proportional to current rating. One important point is that the silicon slice is very brittle and is easily broken in handling. Some types of silicon cell are provided with a protective case and cost slightly more. If you propose to wire up a fairly large array of cells it is probably more economical to use the cheaper unprotected kind and enclose them in a single glass-topped panel.

Normally, you will need to employ at least two cells in series, for there are few circuits that will operate on 0.5v. Cells may be joined in series to obtain whatever emf is required. Most of the circuits in this book require either 3 cells, giving 1.5v, or 6 cells giving 3v. If it is thought that lighting conditions will be less bright than average, an additional cell or two can be added to compensate.

Cells may also be joined in parallel, so as to make extra current available. A series-parallel arrangement allows both increased voltage and current. Large arrays of cells are frequently used to power space-craft and satellites. They have also been used on a solar-powered aeroplane. However, the capital outlay required to build a large array puts this approach beyond the pocket of the average bobbyist. Also there is the practical problem of where to place the array. A satellite can very conveniently have large panels of solar cells projecting into space, and there is plenty of room for cells on the wings of an aeroplane, but the space on top of a portable radio set is severely limited. Consequently we shall use only small arrays for the projects to be described in this book. As examples, Table 1 gives the output from an array of 6 small cells connected in series and tested on a sunny day in autumn. The circuit is shown in Fig. 3(a). Currents of a few tens of milliamperes can

TABLE 1

Output from 6 solar cells in series

Conditions	Load (Ω)	emf (V)	Current (mA)	Power (mW)
Direct illumination	o.c.	3.00	—	—
	330	2.85	8	23
	33	2.25	68	153
Fresnel lens	o.c.	3.10	—	—
	330	3.05	9	27
	33	2.50	76	190

o.c. = open circuit

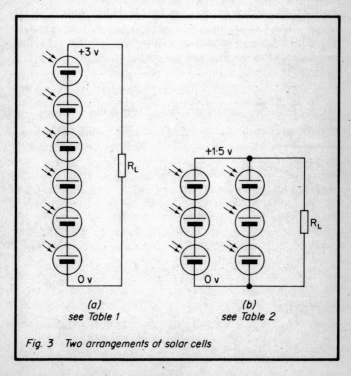

Fig. 3 Two arrangements of solar cells

readily be obtained with no significant drop in emf. Since many of the projects are designed to operate in the *microwatt* range, six series-wired cells are adequate even if the Sun is covered by cloud. The lower half of the table illustrates the effects of concentrating the Sun's rays by using a lens. The lens used was the Fresnel lens-plate from an overhead projector. These are generally about 20cm square, and consist of a sheet of transparent plastic impressed in narrow concentric zones to form a Fresnel lens. Such lenses are ideal for the purpose and are often available in slightly flawed condition from surplus dealers for about £1. The use of a lens will help obtain maximum power from a small array of cells. It can be useful on semipermanent installations, though it is often too cumbersome for portable equipment. It will be noticed that the lens does most good in maintaining emf when current drain is high.

Table 2 shows the output from 6 cells, wired as in Fig. 3b. Such a configuration produces only 1.5v, but allows currents of 40 − 50mA to be drawn without appreciable drop in emf. There is obviously a good reserve of power, since the Fresnel lens makes little difference to the output.

TABLE 2

Output from two parallel sets of 3 Cells in series

Conditions	Load (Ω)	emf (V)	Current (mA)	Power (mW)
Direct illumination	o.c.	1.50	—	—
	330	1.50	4.5	7
	33	1.45	44	64
Fresnel lens	o.c.	1.51	—	—
	330	1.50	4.5	7
	33	1.50	45	68

o.c. = open circuit

The Economics of Solar Cells
In order to keep costs reasonably low and to obviate the need for large arrays of cells, the projects are designed to operate on low voltages (usually 1.5v or 3v) and with low currents (often

less than 1mA). The projects can equally well be powered by one or two dry cells. Readers may prefer to power their devices from dry cells or rechargeable cells and not use solar cells at all! However, there is obvious novelty appeal in having a solar-powered radio or timer and, for the cost-conscious reader, there is also the bonus of getting a free supply of energy from the Sun.

When you are buying the components for a project, the cost of the cells is a major item. Is solar energy the most economical way of obtaining power? It is difficult to make exact comparisons between power sources, for different types of source may operate at different ratings, but in Table 3 a series of comparisons is made. It is assumed that we wish to obtain 1 watt of power by each of the four sources listed. Costs are averaged over various different types and sizes of solar cells, rechargeable cells and dry cells from various suppliers. The installation costs for a solar panel make allowance for mounting and protecting the cells. After installation there are no further

TABLE 3

Comparative costs of different power sources, per watt

Source	Installation (£)	Running cost per hour (p)	Days of running after which cost exceeds solar cells
Solar cells	70	nil	—
Rechargeable nickel-cadmium cells	34	0.01	150
Rechargeable lead-acid cells	30	0.01	167
Alkaline and Leclanche dry cells	nil	8.5	34
Mains d.c. power pack	7	0.005	525

operating costs. Note that the figure quoted provides 1 watt, whereas the projects of this book operate on a fraction of this amount. The installation costs for nickel-cadmium cells include the cost of a constant-current charger. The running costs include replacement costs of the cells after 600 charge-discharge cycles. Some types of cells can not be recharged as many times as this. Lead-acid cells can deliver high currents, so the actual cost of cells per watt is very low if the cells are operated as maximum current. As a result of this the costs of installation have worked out lower than if the power is to be used at more modest rate. The mains-powered pack is the cheapest to install and is economical in running costs, but is not suitable for portable equipment or for equipment that is situated far from a mains socket. It is clear from the table that solar cells sooner or later begin to pay their way. This advantage of solar cells is likely to increase in future as more efficient yet cheaper cells become available. Their main component, silicon, is the fifth most abundant element in the Universe, and the second most abundant in Earth's rocks so it is unlikely to become scarce. On the other hand, the cost of rechargeable or dry cells is likely to increase steadily, as is the cost of the electricity needed to power the battery charger. In addition, solar cells have the advantage that no maintenance is required. A solar-powered device can continue working for years without attention. There is no need for recharging or renewing batteries and no chance of power failure (during daylight hours, at least!).

The one serious problem with solar cells is their dependence upon having a reasonable level of illumination. This may not matter for devices that need to be operated only by day (such as the greenhouse alarm, p.65). In other projects we overcome the problem by using a rechargeable cell to keep the circuit running during darkness. This puts installation costs up to £84 *per watt* (an extra pound or two for the cost of eventual cell replacement), but there is still an overall saving. Whether the hard economic facts are of importance to you or not there still remains the fact that solar-powered projects are a challenge to design and are fun to build. It is hoped that the reader will enjoy both the fun and the challenge while trying out some of the projects that follow.

CHAPTER 3

Power Supplies

In this chapter we describe various projects concerned with generating electricity by solar power. We begin with power packs to be used with the circuits described in the following chapters. The chapter finishes with two battery-charger circuits. These can be used for charging batteries used in calculators, portable radio sets and other equipment, provided that they are of the rechargeable type. Sometimes a device may be *indirectly* solar-powered by driving it from rechargeable cells that are recharged by a solar battery-charger. This approach may be preferred if it is not convenient to mount an array of cells on the device itself, or if it is normally operated in a low-light environment.

Basic Power Supplies

These consist of a number of cells wired all in series or in a series-parallel arrangement (Fig. 4). The numbers and arrangement depend upon the requirements of the circuit that is to be powered and the amount of current that can be generated by each cell. These requirements are given at the beginning of the description of each project.

The method of mounting depends on the type of cell purchased. The more expensive encapsulated types generally have terminal pins at the rear. Such cells can be mounted on a sheet of thin plywood, fibre-glass board or Formica (Fig. 5). They may be secured by Araldite or other adhesive, or by double-sided self-adhesive pads (e.g. "sticky fixers"). Holes are bored in the board to allow the pins to pass through to the back of the board, where the wiring is done. The cells will not need further protection against mechanical damage but, if there is a likelihood of their being exposed to rain, a water-proof cover must be provided. A simple way of doing this is shown in Fig. 5.

The cheaper cells may be already provided with soldered flying leads (Fig. 6a) or they may have none (Fig. 6b). The first thing to do is to mount them on a firm support of the type described above. If holes are drilled in the board, the flying

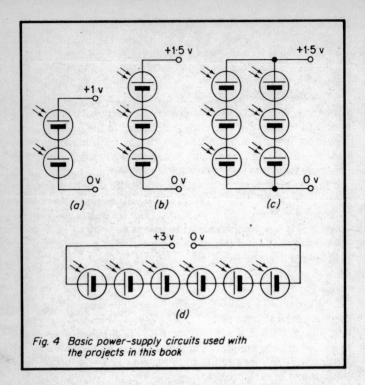

Fig. 4 Basic power-supply circuits used with the projects in this book

leads may be taken through to the rear of the board, where connections are made. This will give the panel a neater appearance. When mounting the cells, take care not to press them too hard, or they may crack.

If your cells have no leads, drill large holes in the board where each cell is to be mounted, so that there is access to the metallic coating on the back surface of the cell (Fig. 6b). It is an easy matter to solder wires to the metallic coating, and to the strips on the front surface of each cell. Some firms supply an array of cells already mounted, leaving you to connect them as you wish.

Should a cell become cracked, it is usually not too difficult to repair it. Place the pieces front-surface-down on a flat surface and assemble them in their correct relative positions. The metal coating on the rear surface will probably be cracked; bridge

10

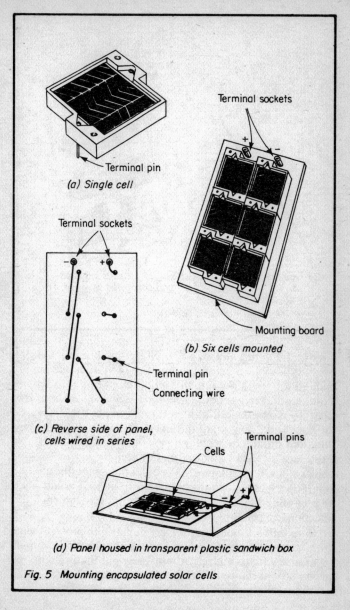

(a) Single cell

Terminal pin

Terminal sockets

Mounting board

(b) Six cells mounted

Terminal sockets

Terminal pin

Connecting wire

(c) Reverse side of panel, cells wired in series

Terminal pins

Cells

(d) Panel housed in transparent plastic sandwich box

Fig. 5   Mounting encapsulated solar cells

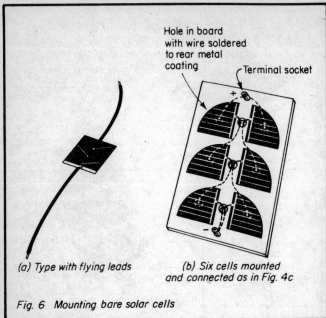

Hole in board with wire soldered to rear metal coating

Terminal socket

(a) Type with flying leads

(b) Six cells mounted and connected as in Fig. 4c

Fig. 6   *Mounting bare solar cells*

every crack with a blob of solder. Keep the solder blobs as small as possible, otherwise this will lead to difficulties in mounting later. Make sure that *all* the fragments are inter-connected. Apply adhesive liberally to the rear surface of the cell. Turn over the cell with great care and place it in a position on the board. Now repair the breaks in the metallic strips on the front surface, by running a blob of solder across each. Make sure *all* breaks are mended, but keep the size of the solder blobs to a minimum. Do not let the solder spread unnecessarily, for this will obscure the front surface and make the cell less efficient. A cell repaired in this way should function in a perfectly normal manner.

Once cells are mounted it is very desirable to protect them from damage. A sheet of glass, perspex or polystyrene carried on a narrow frame is ideal. Alternatively put the array in a plastic box, as shown in Fig. 5.

The array is normally mounted on or near the device that

it is powering. To obtain maximum power, the Sun's rays should strike the array perpendicularly. In practise this is usually not important. A horizontally mounted array receives sufficient illumination during the whole day except early morning and late evening. If maximum power is required, a mounting such as that of Fig. 7 allows the Sun to be tracked throughout the day. The Fresnel lens concentrates a beam of light on to the area covered by the cells. For this to be most effective, the cells must be mounted as close together as possible, so that the beam may be concentrated on a small area.

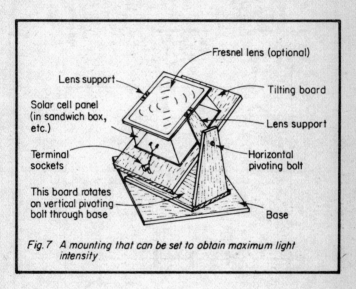

Fig. 7 A mounting that can be set to obtain maximum light intensity

"Solapak"

Fig. 8 shows the circuit for a general-purpose power pack suitable for the experimenter. It is a handy item to have when building any of the projects in this book. In its simplest form it consists of 6 cells with a means of switching them all in series to give a 3v output or in the series-parallel arrangement to give 1.5v. Eight cells can be similarly wired to give 4v or 2v.

The voltage multiplier is optional but is a very useful part of this project. Although it is possible to obtain, say, 6v by

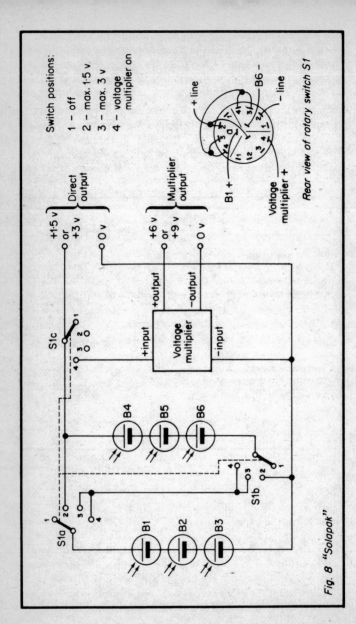

Switch positions:

1 – off
2 – max. 1·5 v
3 – max. 3 v
4 – voltage multiplier on

Direct output

+1·5 v or +3 v

0 v

Multiplier output

+6 v or +9 v

0 v

+output

−output

Voltage multiplier

+input

−input

S1c

B4 B5 B6

S1b

S1a

B1 B2 B3

+ line

B6 −

− line

B1 +

Voltage multiplier +

Rear view of rotary switch S1

Fig. 8 "Solapak"

14

using 12 cells in series, this may be an unwarranted expense and will require a double-sized panel to carry the cells. Although several projects require high *voltage* (6v is high by solar-cell standards) the *current* requirements may be less than 1mA. It is uneconomic to provide 6v at 50mA when only 1mA is to be used.

The voltage-doubler shown in Fig. 9 requires very few components and is easy to construct on a scrap of circuit-board. The heart of the circuit is the oscillator, which has a frequency of about 1kHz, and is built from half of a 4001 CMOS IC containing four 2-input NOR gates. We use CMOS a lot in this book, for it has the twin advantages of low current consumption and the ability to operate on all voltages between 3v and 15v. Note that if the supply voltage falls *below* 3v, the IC may cease to operate, so good exposure to sunshine (or failing that, an extra solar cell) are essential.

The output from the oscillator controls a second IC, a CMOS quadruple bilateral switch. When the output of the oscillator is high, switch A connects point X to +V, switch D connects point Y to 0v. Switches B and C are off. When the oscillator goes low, switch B connects X to 0v and switch C connects Y to +V. Switches A and D are off. Thus points X and Y alternate in polarity. They are connected to an arrangement of diodes and capacitors that acts as a voltage doubler. When X is positive to Y, C1 charges to +V, through D1. When Y is positive to X, C2 charges to +V through D2. Each capacitor is charged to +V, so pd across the pair of them is twice +V. With an input voltage of 3v, this doubler gives 5.5v when its output in open-circuit. This is sufficient to operate most IC's rated at 6v, though, if a 7 cell power supply is used (3.5v), there is more leeway. When CMOS is operated on a 15v supply the resistance of each switch is only 80Ω, but this resistance increases as supply voltage decreases. Consequently the voltage doubler has high output impedance. It will deliver a current of 100–200µA with a drop of output voltage of less than 1v, but can not supply greater currents. A hundred microamperes is all that is needed for some projects and the simplicity of construction and small physical size of this circuit are advantages. The current output can be doubled by wiring a second 4066 IC in parallel with IC2, pin-for-pin,

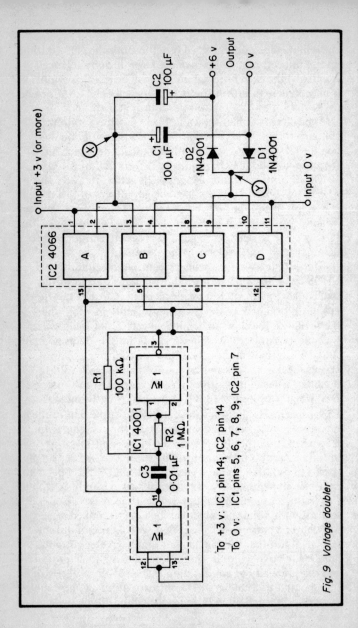

Fig. 9 Voltage doubler

16

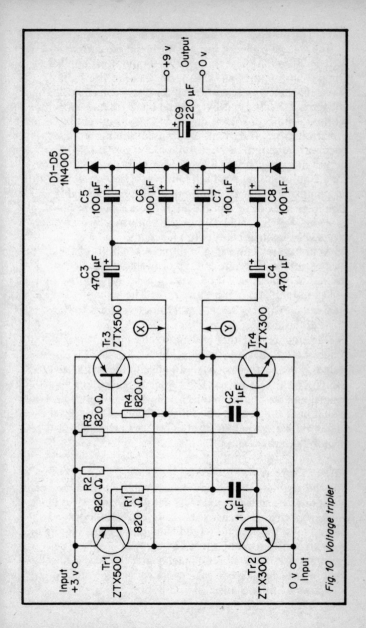

Fig. 10 Voltage tripler

17

and taking its outputs to points X and Y.

The voltage tripler of Fig. 10 provides more current at higher voltage. Against this must be set the larger number of components required and the much greater physical size. When connected to a 3v solar cell pack, its open-circuit output voltage is 8.2v. It can deliver several hundred microamperes with no significant voltage drop. At 2mA its output is 7v, and at 6mA it is 6v. The circuit can be used with higher input voltages if required. A 7-cell input (3.5v) gives a full 10.5v output dropping only to 8.8v with 7mA output, and to 6.6v for a 20mA output. An 8-cell input (4v) gives 12v output, dropping to 7.6v for a 23mA output. It is also allowable to use inputs at 5v or 6v, but the latter figure should be be exceeded, for this could damage the transistors by excessively reverse-biasing their base-emitter junctions.

As in the previous voltage multiplier, the heart of the circuit is an oscillator. This one consists of two cross-coupled relaxation oscillators. At one instant point X is positive of Y and an instant later Y is positive of X. The network of capacitors creates a potential difference that causes a current to flow through the chain of diodes.

Construction of either of these voltage multipliers is a simple matter. Begin by building the oscillator. This can be tested by connecting a crystal earphone to the 0v line and to points X or Y, through a $0.1\mu F$ capacitor. In each case a distinct audible tone should be heard. The multiplying network is built next and connected to the oscillator. Take special care to observe correct polarities, both for the diodes and for the electolytic capacitors.

Battery Chargers

The chargers described in this chapter may be built as complete units, or as attachments to the "Solapak" circuit. Most kinds of equipment that use rechargeable cells (e.g. calculators and radio sets) work from nickel-cadmium cells. These require to be charged at constant current. By contrast, lead-acid cells require constant voltage. Commonly-available lead-acid cells require high charging currents, so are beyond the scope of a low-powered solar charger.

Constant Current Charger A

This provides up to 10mA charging current. It is suitable for recharging small nickel-cadmium cells such as button-cells, the small batteries made for mounting on printed circuit boards and cells of PP3 and AAA (U16) sizes.

It requires a power supply of six solar cells in series (+3v). The circuit makes use of a constant-current IC, the 334Z (Fig. 11). The value of the current is set by VR1. The required value can be calculated by the formula:

$$R = 67.7/I \text{ ohms}$$

in which R is the set value of VR1, in ohms and I is the required current, in *milliamperes*. If R1 has maximum resistance 100Ω, the minimum current obtainable is 0.677mA. Most button cells are charged at 1.5 − 3.0mA. The maximum current obtainable from this IC is 10mA. When the circuit is assembled, a millimeter may be connected to the load terminals. Then VR1 is adjusted to give various current values between 1mA and 10mA and at each setting the position of the pointer-knob of VR1 is marked on a scale. This procedure calibrates the charger for a variety of cells. If you have only one type of cell to charge, a fixed-value carbon resistor of suitable value may be substituted for VR1. The voltage-drop across the IC is about 1v. This does not allow more than one cell (up to 1.25v) to be charged at one time if the solar cells are providing only 3v. To charge

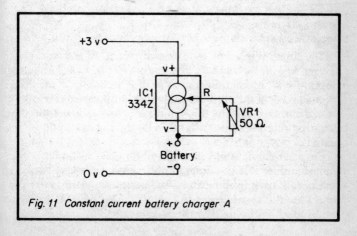

Fig. 11 Constant current battery charger A

more than one cell simultaneously, all that is needed is to increase the input voltage. For example, to charge 4 cells in series, we first calculate their total emf. Four times 1.25v gives 5v. Next add 1v for the drop across the IC. This gives a total of 6v, which is the minimum required to operate the circuit. If the input voltage is less than this, the drop across the IC is less than 1v, which is insufficient to turn it on. The input voltage of 6v can be obtained either by using 12 solar cells in series or, preferably, by using 6 solar cells and the voltage trebler. Note that applying *excess* voltage to this circuit has no effect on the charging current, provided that the maximum forward voltage, 30v, is not exceeded.

Should the level of sunlight fall while a cell is being charged, so that the IC is turned off, a small leakage current flows back through the IC from the battery. This could be prevented by putting a diode between VR1 and the positive terminal of the battery. However, the leakage is less than $10\mu A$ so it is very doubtful if this addition is worth while.

Constant Current Charger B

All types of nickel-cadmium cell may be charged by using the circuit described above, but, since its maximum rate is 10mA, it could take up to 50 hours to recharge an AA(HP7) size cell. The circuit of charger B provides constant current up to 50mA. This is sufficient to charge an AA(HP7) cell in about 12 hours. The circuit can be adapted to run at higher rates, charging larger cells, provided that sufficient current is available from the solar panel.

The circuit is based on the constant current IC used for charger A, with the addition of a transistor (Fig. 12). The constant current I is set by choosing the value of R according to the formula given in the previous section. A small current flows from the base of Tr1 and through the IC. A larger current flows from the collector of Tr1 directly through R1 to the battery.

For a current of 50mA the value of R1 must be 1.35Ω. Variable resistors of this order of resistance are not easy to obtain, so R1 is a fixed resistor. The nearest standard values for R1 are 1.2Ω (giving 56mA) or 1.5Ω (giving 45mA). Either one of these can be used, or two resistors of 2.7Ω can be wired

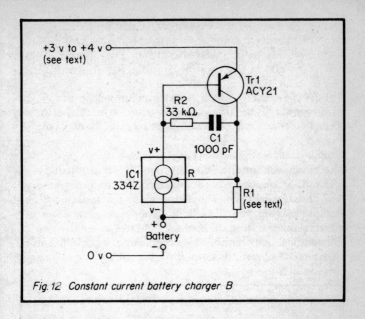

Fig. 12 Constant current battery charger B

in parallel to obtain 1.35Ω. If a number of different settings are required, use a rotary switch to bring one of a selection of different resistors into circuit.

A six-cell solar panel will not be able to provide 50mA without loss of voltage if the cells are small in area. If the cells are large there will be no problem but, if they are small, the extra current may be provided by wiring 8 cells in parallel to give a 4v open-circuit output that drops to 3v when connected to the charger. As explained in the previous section, a higher voltage is required if more than one cell is to be charged at one time. With charger B we can not make use of a voltage multiplier, for this type of circuit will not supply as much as 50mA from solar cells. To increase voltage, add further solar cells in series at the rate of 3 or 4 solar cells for each cell to be charged.

CHAPTER 4

Radio Receivers

The ZN414 radio IC, which operates on low voltage and consumes very little current is easily powered by solar cells. This chapter includes three designs of increasing complexity.

Simple Headphone Radio Set

This circuit will operate well using only 3 solar cells wired in series. It is suggested that it could be built into a small case, with the solar cells on top and attached to the band of the headphones (Fig. 13). This "radio headphone" can be worn while relaxing in the garden, or used indoors under artifical illumination. Alternatively it can be made as a portable unit with a single earpiece or ear-plug for listening.

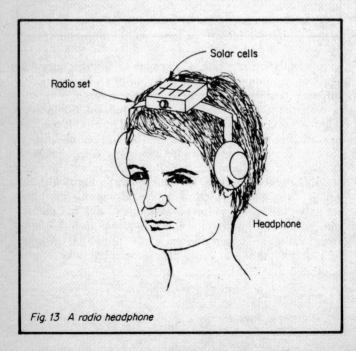

Fig. 13 A radio headphone

The circuit is shown in Fig. 14. To make the tuning coil L1, you need a length of ferrite rod 5 or 6 cm long and 9.5mm in diameter. Wrap a piece of writing paper around the central region of this and fix it with Sellotape. Then wind a coil consisting of 85 turns of 40 swg enamelled copper wire. Adjacent turns should touch. Secure the ends of the coil to the rod by spots of Bostik or Araldite. If the set is *not* to be used for a "radio-headphone", a longer ferrite rod, up to say 12cm could be used to give improved reception.

Variable capacitor VC1 is a tuning capacitor with air or plastic dielectric. It can be of the rotating vane type or the pressure type, though the rotating vane type is preferred. The terminal to which the rotating vanes are connected should be wired to the R1–C1 junction, the other terminal going to the input of IC1. The decoupling capacitor C2 should be connected as close as possible to the Earth and Output pins of IC1. Any high impedance earphone may be used; if you are making a "radio headphone", wire the two phones in parallel. Their combined resistance must be 250Ω or more.

With careful arrangement of components, this circuit requires very little space. The supply voltage should be between 1v and 1.5v, preferably towards the upper end of that range.

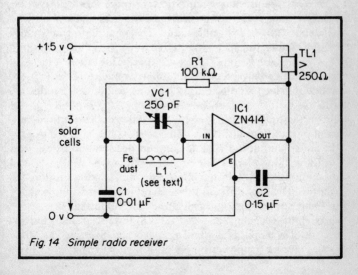

Fig. 14 Simple radio receiver

23

Three solar cells will provide this easily, even on fairly dull days. One point to remember is that the power supply to the circuit *must* be switched off while any soldering operations are being performed. If this precaution is not observed there is high risk of destroying the IC.

The spindle of VC1 is to be fitted with a pointer knob, so that settings for different stations may be marked on the case. The ferrite aerial is directional; reception from the weaker stations can be improved by turning the aerial (or your head!) until the best result is obtained. With this set it is possible to receive all the medium-wave stations of the BBC and local radio. At nights many European stations are clearly received. If the range over which the set can be tuned does not cover all the BBC stations, try adding a few more turns to L1, or try removing some turns.

Headphone Radio with Amplifier

This circuit (Fig. 15) is rather more complex than the one described above, but provides an output of markedly greater volume and quality. Since the basic radio circuit is the same for both, it is easy to upgrade the simple radio circuit by adding the amplifying section of this circuit.

As with the previous circuit, power from 3 solar cells is sufficient, and it can be constructed either as a "radio-headphone" or as a portable radio. The amplifier is a new type of CMOS operational amplifier IC that has the valuable feature of working with a supply of only ± 0.5v. This makes it ideal for use in solar cell projects. With a supply voltage of 1.5v it is well within its operating range. Its other perhaps even more valuable feature is its low power consumption. If pin 8 is connected to the positive supply, the quiescent current requirement of the IC is only 10µA. This restricts the band-width of the amplifier, but this is of no consequence at audio frequencies.

The radio section of this circuit, including the winding of the tuning coil, is almost as already described in the previous section. The only differences are that the circuit is connected to the positive supply through a resistor (R2) instead of through the earphone, and that C2 is of lower capacitance than before. The amplifier is wired as a high-gain inverting amplifier. If we consider the power-supply rails to be +0.75v and −0.75v,

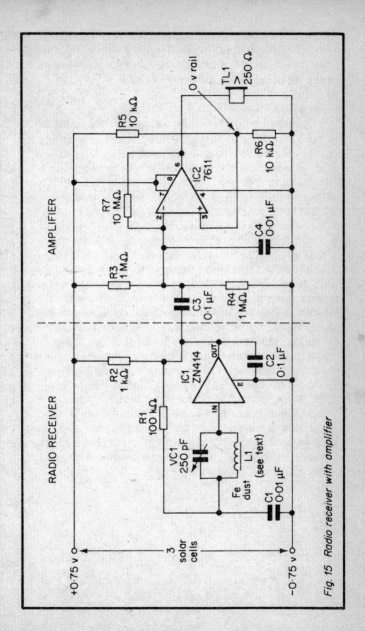

Fig. 15 Radio receiver with amplifier

25

we obtain 0v by the potential divider, R5—R6. The non-inverting input is tied to 0v. The inverting input is also held at 0v, by the potential divider R3—R4, but is coupled to the output of the radio circuit by way of C3.

Personal Radio Set

For those who prefer not to wear headphones, this table radio set provides ample volume for listening under quiet conditions. Like the other radio sets, it is powered by only three solar cells. Indoors, it can be placed beneath a table-lamp with a 100 watt bulb. The first section of the circuit is identical with the corresponding section of the radio sets described above. The details for winding the coil are as given on p.23.

The set has two stages of amplification, both consisting of operational amplifiers. The two op amps are contained in a single IC, the 7622. This is the dual version of the 7611 used in the circuit of Fig. 15, and shares with it the twin advantages of low operating voltage and low current requirements. The amplifiers are connected in the non-inverting mode and capacitor-coupling is used between stages. In order to obtain the maximum possible volume the loudspeaker is connected to the −0.75v rail. At full volume this results in clipping of the wave-form. Distortion occurs when strong stations are being received. Potentiometer VR1 acts as a volume control. By adjusting VR1, volume can be reduced just sufficiently to eliminate distortion. The circuit is shown in Fig. 16.

Although this circuit is intended for solar power, a 1.5v dry cell or a rechargeable cell may be used instead. This whole circuit with its single cell occupies very small volume and is ideal for a "mini-radio". If a 1.25v rechargeable cell is used, resistor R1 may be reduced to about 270Ω.

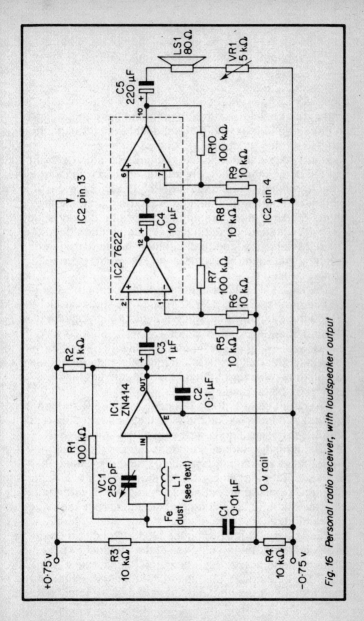

Fig. 16 Personal radio receiver, with loudspeaker output

27

CHAPTER 5

Lamps, Buzzers and Alarms

Door-Key Light

This circuit provides just over 2 watts of illumination, which is quite enough to help you find your door-key in the dark. It can also provide current for a door alert buzzer.

It may seem a strange idea to be using solar power at night but this device stores energy by day for use during the following evening. All day the solar cells deliver a steady charge to the battery at about 9mA. During an average day the cell gains about 70mAh of charge. During the evening the battery is used to power one or two filament lamps requiring about 0.6A. At this rate the battery loses its daily charge in approximately 7 minutes. Something must be done to eke out the stored power.

This circuit (Fig. 17) uses a monostable to turn off the light automatically after 15–20 seconds. The light can be used 20–30 times each evening, which is more than enough for most circumstances. Also, since the battery stores about 100mA, there is a reserve of power for busy evenings. The door buzzer takes only a few milliamperes and is usually sounded for only a second or two at a time, so it does not need to be taken into account in the calculations.

The circuit requires 10 to 12 small solar cells. Only 9mA is required so there is no point in using the more expensive cells with large area and large current production. Alternatively, the circuit can be powered by 6 solar cells and the voltage — tripler circuit (p.17), a cheaper though more elaborate course of action. The constant current IC is wired as explained in Chapter 3 to charge the battery at 9mA. When day-light fails, charging ceases. It is, of course, possible to operate the buzzer by day or night.

The monostable is based on the CMOS 4098 dual monostable multivibrator IC, though only one of the monostables is used. This is wired to be triggered when its +TR input (pin 4) is made high. The output from this monostable (pin 6) goes high and stays high for a period of time determined by the value of R2

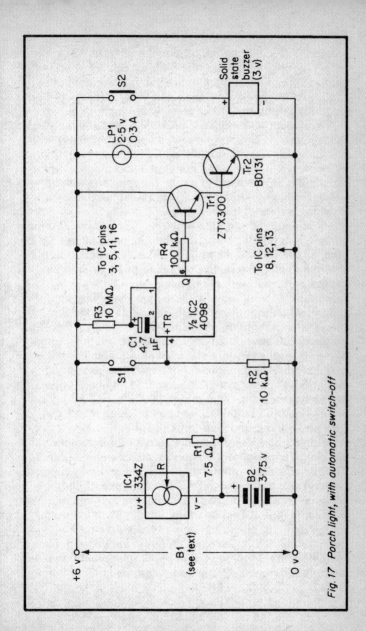

Fig. 17 Porch light, with automatic switch-off

29

and C1. With the values indicated the period is about 20 seconds. For longer periods increase the values of R2 or both. The output from the monostable is only a few milli-amperes. This output is fed to Tr1 and Tr2 connected as a Darlington Pair to provide the necessary gain. Tr2 is a power transistor able to carry 1A or more without damage.

The amount of illumination provided depends on the number of lamps used. For many purposes a single 0.3A lamp is sufficient, especially if mounted with a reflector to direct the light to where it is required. For greater illumination, two or three lamps may be wired in series. The drain on the battery will be doubled or trebled but this will be of no consequence if the lights are used less frequently.

Construction presents few problems. The battery B2 is the compact pcb type, and, as such, can be mounted along with the other components on the pc board. The multivibrator section should be built and tested before connecting it to Tr1. The capacitor C1 must be of the tantalum type. To allow the monostable to work properly the unused inputs must be connected to the positive or negative rails, as listed in Fig. 17, before attempting to test it. Switch S1 is normally a push-button or bell-push placed at some easily located point in the porch. Normally the output at pin 6 of IC2 should be low (0v), but when S1 is pressed the output should go high and stay high for approximately 20 seconds. This sounds a short time, but is more than enough time in which to find a key and unlock the door. In practice you may prefer to reduce the time by decreasing the value of C1 to $1\mu F$.

Switch S2 and the buzzer are optional additions to the circuit. S2 is an ordinary bell-push mounted deside the door. The buzzer is a solid-state buzzer. Obtain one rated to operate on 3v. These buzzers produce a loud sound though they consume little current. To obtain the maximum volume, it is essential to mount them on a firm surface. When connecting the buzzer to the circuit take care to observe the correct polarity. Usually the buzzers are already provided with two leads, one red (+) and one black (−). As an alternative to the buzzer you can use the alarm circuit described later in this chapter.

Flashing Beacon

In low-powered projects such as these we are forced to consider all possible ways of economising in the use of energy. The project just described allowed a reasonably bright light for a limited number of short periods. In this project the lamp is flashed for very short periods indeed, with long gaps between, during which virtually no energy is consumed. Such a flashing light can have many uses as a beacon. This is the basis of the flashing lamps that act as the attention-catchers we often see lining the roads when repairs are being carried out. These are battery-powered and the fact that they are flashing with high intensity but for extremely short periods means that they are very economical of power. Yet a flashing lamp catches much more attention than one that shines continuously. A flashing lamp makes maximum use of limited power.

This circuit (Fig. 18) is based on a special "oscillator-flasher" IC, the LM3909, and needs the minimum of external components. It flashes the light-emitting diode at a rate slightly less than 1 Hz. Its chief advantage is that the *average* current consumed is only 0.32mA. A single nickel-cadmium button-cell is all that is required to power the circuit and its charge is easily kept topped up by 6 small solar cells in conjunction with

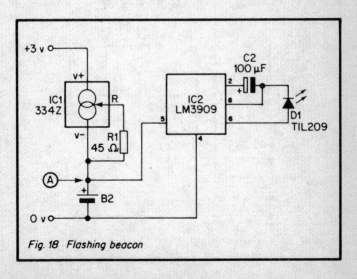

Fig. 18 Flashing beacon

the constant-current IC. The maximum charge rate for most button cells is 1.5mA. In a period of 24 hours the circuit needs only 5 hours exposure to bright light (sunlight or artificial light) to keep it fully topped up. The cells contain sufficient reserve of charge to run the circuit for many days without illumination. Such a circuit is perhaps more economical than it needs be. The circuit of Fig. 19 flashes 4 LEDs simultaneously. It runs from a single button cell, with three solar cells to recharge it. For each hour of use it requires one hour of exposure to light to recharge it.

Construction of these circuits presents no problems. The current-setting resistor R1, produces charging current of 1.5mA. It may be obtained by wiring a 33Ω resistor in series with a 12Ω resistor. The LEDs can be of any type, though jumbo-sized red LEDs seem to produce the most striking effect.

The circuits of Figs. 18 and 19 are intended for a self-contained unit that is to be used at night and recharged by

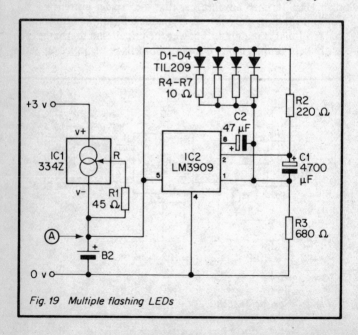

Fig. 19 Multiple flashing LEDs

day. Normally, it would not be suitable for LEDs to be powered directly by solar cells, since LEDs are not easily visible in sunlight or in bright room lighting. However, if the LEDs are to work in a relatively low-lit situation indoors, the cells can be located outdoors or by a window. Three small solar cells are required, connected in series. They are joined to the 0v rail and to point A (Figs. 18 and 19). IC1, R1 and B2 are omitted. Such a system could be used as a warning indicator; for example, a reminder that a door essential to the security of the house has been left open.

Touch-operated Lamp

This too can be used as a porch light, or it could light a particularly dark cupboard, or be used as a bedside lamp for a child or invalid. The lamp is turned on by a touch-switch, which is easy to find in the dark.

As Fig. 20 shows, the lamp is powered by a battery of 3 nickel-cadmium cells, recharged from solar cells by day. The

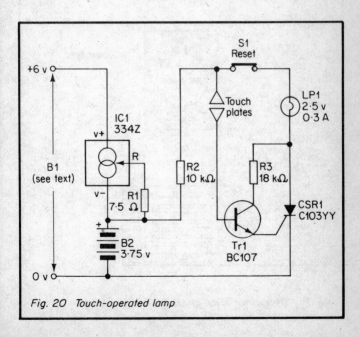

Fig. 20   Touch-operated lamp

33

switching device is the thyristor, CSR1. When the circuit is first connected to the battery no current flows through the thyristor so the lamp is off. The touch plates are two metal plates with a narrow gap (about 1mm) between them. Fig. 21 shows some designs for touch plates; the simplest method is to use strip-board as shown at (a) but, for a more professional appearance, the contact areas can be prepared by etching printed circuit board with one of the designs (b) or (c). When

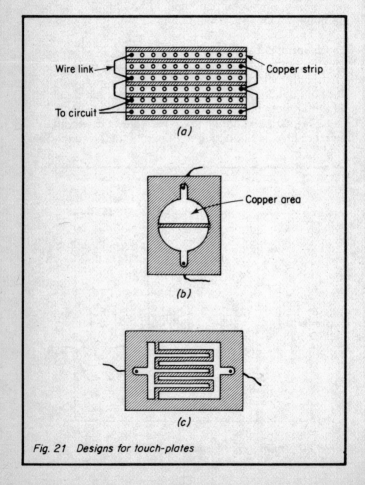

Fig. 21   Designs for touch-plates

a finger is placed so as to bridge the gap between the plates, a minute current flows through the flesh, and through R2 to the base of Tr1. This turns on Tr1, effectively connecting the gate of the thyristor to the positive line. This positive-going pulse turns on CSR1, making it conduct so that a large current flows through LP1. The lamp remains lit until push-button S1 is pressed. This is a *push-to-break* button. When it is pressed, the flow of current is interrupted. Once current has ceased to flow, CSR1 does not resume conducting, even when S1 is released. When using this circuit, one must remember to conserve power by using the reset button promptly. This circuit can be put to several other uses, for its essential feature is that it is triggered by a flow of current between the two plates. On washing-day it can function as a rain alarm. Contact plates like those at (a) and (c) in Fig. 21 make the most suitable sensor. The sensor is placed outdoors, contacts facing upward. As the first few drops of rain fall the gap between the surfaces is bridged and the circuit is triggered. The remainder of the circuit is indoors with the lamp in some prominent spot in the kitchen or living room. As the rain begins the lamp lights, warning that the washing should be taken in as soon as possible. Another use for this circuit is as a water level detector. It can show a warning light when a tank of water is about to overflow. In this application a sensor plate similar to Fig. 21(b) is placed inside the tank, so that the gap between the areas of copper is at the highest allowable level.

Door Buzzer

A door buzzer circuit such as that shown in Fig. 22 can be used on its own or to replace the solid-state buzzer shown in Figure 17. For daytime use it does not require a battery and charger. It can be powered by six medium-sized solar-cells, since it requires only 30–40mA. The circuit is based on a single CMOS IC, the 4001 quadruple 2-input NOR IC. Only two of the four gates are used, the inputs of the unused gates being connected to +3v to reduce power requirements and to ensure the correct functioning of the other gates. The gates used in the circuit have their inputs connected so, in effect, they are simple INVERT or NOR gates. When connected

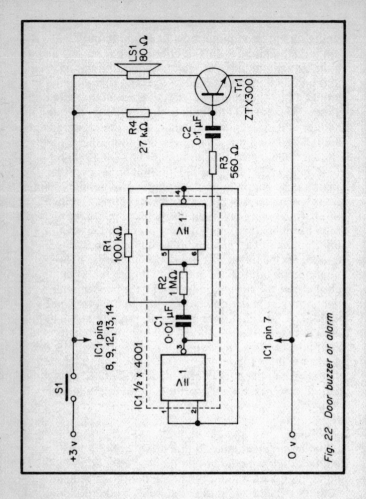

Fig. 22 Door buzzer or alarm

as shown they act as an astable multivibrator. This has two states in each of which the output of one gate is high and the output of the other is low. The circuit oscillates at about 1kHz. The output of gate 3 supplies base current to Tr1. Pulses of collector current are generated at 1kHz, causing a note of the same frequency to be emitted from the loudspeaker. The construction of this circuit is very straightforward, and it can

36

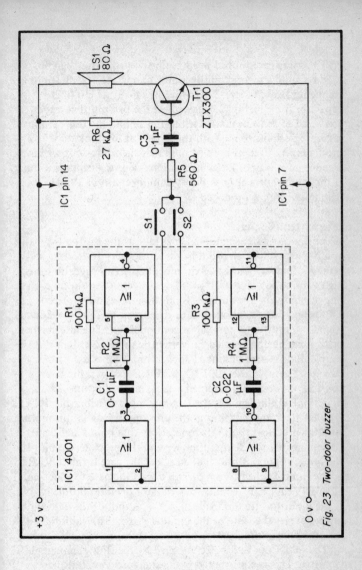

Fig. 23 Two-door buzzer

be compactly assembled on a very small scrap of circuit board. If it is desired to alter the pitch of the note, either C1 or R1 may be altered. Increasing C1 or R1 lowers the pitch. If R1

37

is altered appreciably, alter R2 to about 10 times the new value of R1.

The spare gates in IC1 need not be wasted. They can be used to build a second multivibrator to operate at a different frequency, say 500Hz. To do this we substitute a 0.022μF capacitor for C1. Fig. 23 shows how the two multivibrators are wired so as to be connected to the amplifying transistor by separate push-buttons. With the button at the back-door producing a different note from that at the front-door we have a convenient way of knowing at which door the caller is waiting. If there are callers at *both* doors simultaneously, a third distinctive note is produced.

Intermittent Buzzer

A continuous note, sounding for as long as the button is pressed, is suitable for door alert but is less satisfactory for alarms in general. In a fairly noisy environment an alarm needs to have high volume to make itself heard. An intermittent note, even one of relatively low volume, is much more readily noticed. This is why the telephone bell or trimphone warbler is made to sound intermittently instead of continuously. The intermittent buzzer can be constructed for very little more cost than the simple door buzzer described above, for it makes use of the two spare gates in the 4001 IC. As Fig. 24 shows, the buzzer consists of two interconnected astable multivibrators. The upper one in the figure is the fast-running multivibrator. R1 and C1 have the same values as in the previous circuit so this multi-vibrator produces the 1kHz tone. Note that pins 5 and 6 are *not* connected together as they were in the previous circuit. If the input at pin 6 is low, the output of the gate is the INVERT of the input applied to pin 5. This allows the gate to function as an INVERT gate, so the multivibrator operates normally. If the input to gate 6 is made high, the output of the gate stays low whatever the state of the input at pin 5. This inhibits the action of the multivibrator.

The inhibiting action is controlled by the slow-running multi-vibrator. This is constructed in the standard way, but the capacitor (C2) has a large value to give the vibrator a period of about 2Hz. The output from pin 10 is high for 1 second than low for 1 second. Whenever the output goes high the fast running

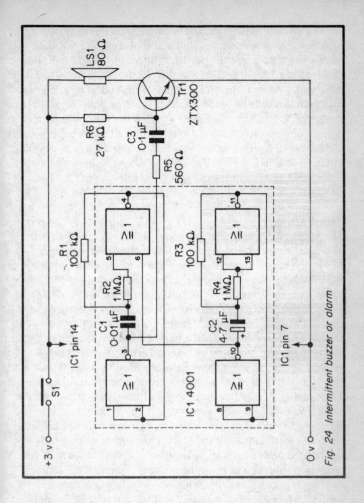

Fig. 24 Intermittent buzzer or alarm

multivibrator is inhibited. The effect is to produce 1-second bursts of tone, with 1-second gaps between them. If a more agitated sound is required, the rate of the slow multivibrator may be increased by reducing C2 to 2.2μF or 1μF. The buzzer is easily constructed on a small piece of circuit board. It is small enough to be mounted in a tobacco-tin or box of similar size. It is important that the capacitor used for C2 should be

of the tantalum bead type, not an aluminium electrolytic
capacitor. First wire up the slow multivibrator and test it by
connecting a voltmeter to pin 10. The needle should indicate
0v for about 1 second, then rise sharply to +3v, remain there
for about 1 second, then fall sharply to 0v, and so on. The fast
multivibrator is to be wired up next, together with the amplifier
transistor Tr1 and the loudspeaker. Test the fast multivibrator
before joining it to the slow multivibrator. If pin 6 is temporarily
connected to the 0v line and power is switched on, a note with
1kHz pitch is heard. Finally the wire between pins 6 and 10 is
soldered in place. On testing again, the intermittent note will
be heard.

Visual Alarm

The human ear is extremely sensitive to sounds of low power.
In a quiet environment, a few milliwatts of sound energy can
be clearly heard. By contrast, the eye is much less sensitive.
A 100W lamp does not appear particularly bright, yet the
sound coming from the 100W amplifier can be almost deafening.
The loudspeakers in the circuits of Figs. 22–24 operate at just
over 0.1W, and give adequate signals. Their output is equivalent
to that of a low-powered torch bulb working with a rather flat
battery! In solar-powered circuits we can not spare power for
bright lights so an alternative approach must be used. This
alarm (Fig. 25) employs a small microammeter as its display.
The timer IC is wired as an astable multivibrator. It produces a
square-wave output with a frequency of approximately 1Hz.
This is fed to the microammeter, causing its needle to flick
back and forth along the scale. This method of display uses
the limited power to produce *motion,* the power by which we
see this motion comes from the ambient illumination – direct
or reflected sunlight or room lights – which is in plentiful
supply. The only circumstance in which this type of alarm is
not seen is in complete darkness.

The IC used for this circuit is the CMOS equivalent of the
well-known 555 timer IC. Since it is a CMOS device, it needs
only a few microamperes to drive it. It is rated to operate over
the range 2v–18v, but it has been found to operate perfectly
well with as little as 1v. In operation, the whole circuit takes
only 250μA. This means that only two of the smallest (and

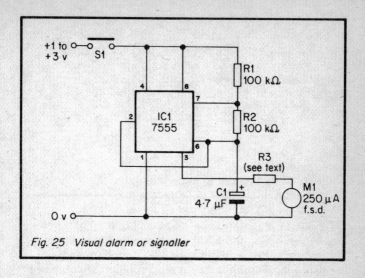

Fig. 25 Visual alarm or signaller

cheapest) type of solar cell are needed to power the circuit. If lighting conditions at the solar cells are likely to be low, it is advisable to use three cells in series instead of only two. Such a simple circuit with an 8-pin IC occupies very small space. The capacitor must be of the tantalum bead type, and this too makes for small size. It is feasible to build the circuit on a piece of circuit board about 2cm square and house it inside the casing of the meter. The prototype used an inexpensive miniature signal-strength meter, rated at $250\mu A$. When operating on a 3v supply, the best value for R3 was found to be $10k\Omega$. The value for other meters and operating voltages can be calculated from the equation

$$R3 = (V - I_{fsd} R_m)/I_{fsd}$$

where V is the operating voltage, I_{fsd} is the full-scale deflection current of the meter and R_m is the resistance of the coil of the meter. In the prototype, R3 came to 11375Ω but to obtain a vigorous alarm signal a resistor at a standard value slightly less than this was chosen. The meter used in the prototype had an attractive green scale with a relatively thick needle. It was found that this produced a sufficiently striking visual effect on

41

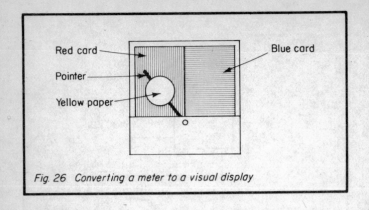

Fig. 26 Converting a meter to a visual display

its own. If the meter you intend to use has a dial of the more conventional type, it can be made to attract more attention by covering the back plate with coloured cardboard (or pvc insulating tape) as in Fig. 26. A disc of coloured paper glued to the pointer helps to enhance the effect.

Attention-catcher

Another circuit allied in function to the one described above is shown in Fig. 27. It actuates a small microammeter in much the same way but has the advantage that it can operate under low light levels, even when powered by only two small solar cells. The circuit uses the same 7611 operational amplifier that was used for the radio circuits of Chapter 4. It readily operates on ± 0.5v and, once started, will continue to operate even when the voltage falls below this. In this application the amplifier is connected as a square-wave generator or astable multivibrator. Suppose that its output (pin 6) has just fallen from high (+0.5v) to low (−0.5v). The potential at its non-inverting input (+) is immediately brought low by means of the feed-back resistor R4. The potential at its inverting input (−) remains high because of the charge on C2. With the potential at the non-inverting input lower than that at the inverting input, the output stays low. However, with low output there is a flow of current from C2 through R3 to pin 6. This gradually discharges the capacitor and the potential at the inverting input gradually falls. It continues to fall toward −0.5v; eventually

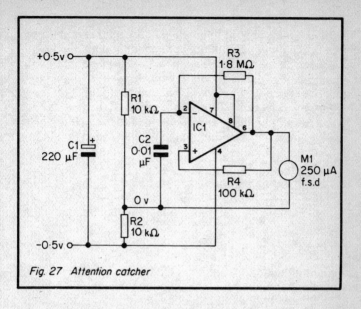

Fig. 27 Attention catcher

it becomes just less than the potential at the non-inverting input, which is slightly less negative than −0.5v. At this point the output swings sharply from −0.5v to +0.5v. Now we have the reverse conditions, with C2 being gradually charged by current flowing from pin 6 and through R3. The potential at the inverting input rises slowly toward +0.5v. Just before it reaches this level it exceeds the potential at the non-inverting input and the circuit reverts to its original state. The IC requires only 10μA, but the switching of 250μA through the meter produces voltage fluctuations in the circuit that are damped by the large-value capacitor C1. This can be of the aluminium electrolytic type. The meter used in the prototype was a low-cost centre-zero tuning-meter. It was found that no resistor was required in series with this meter. With other meters a resistor may be required to prevent excessive swing of the needle.

This device was designed as an attention-catcher rather than an alarm. It can be built as a small unit able to stand anywhere to attract attention. Placed beside the telephone it acts as a reminder to make an important call. On top of the TV set is will remind you to switch on for a special programme. If you are going out and want to leave a note for another member of the family to see when they come in, put the device on top of the note. It will soon attract attention to the note. It needs moderately bright light to start it operating but continues to operate even if the light level is reduced, so it is ideal for leaving for hours under variable lighting conditions. The smoothing capacitor C1 also holds a small reserve of charge that will help bridge any momentary interruptions in illumination.

CHAPTER 6

Bicycle Speedometer

Mechanical speedometers need an appreciable amount of power to drive them. Unless they are very carefully adjusted a certain amount of additional pedal-pushing is required. This circuit (Fig. 28) requires only 6 milliwatts and the whole of this comes from the Sun. The circuit requires only 2mA at 3v, so it can be powered by six small solar cells.

The phototransistor Tr1 detects the sudden pulse of light from a small reflector fixed to the front wheel of the bicycle (Fig. 29). The reflector is a piece of sheet aluminium about 2cm square. It is fixed to a spoke of the wheel and angled so that when it is opposite to the phototransistor it reflects light from the sky directly on to it. The phototransistor is mounted on one tine of the front fork, pointing inward toward the reflector. It must be covered so that it can receive light only from the direction of the reflector but not from other directions. It is better if a black screen is mounted on the opposite tine to block light from objects by the roadside. A certain amount of experimentation is needed to obtain the ideal operating conditons for any given make of bicycle. The 2N5777 is a photo-Darlington device. It is not a simple phototransistor but incorporates an amplifying stage. Consequently a small variation in the amount of light falling on the transistor produces a large variation in the potential at the collector. The operational amplifier IC1 is connected as a differential amplifier. The potential at the non-inverting input is adjusted, by setting VR1, so that it is equal to the potential at the inverting input. This is, of course the same as the collector potential of Tr1. When the reflector passes Tr1, the level of light falling on Tr1 is momentarily increased. This causes a fall of potential at the inverting input, with a consequent rise at the output. The rising output from the amplifier produces a positive-going pulse that passes across C1 and triggers the timer IC. This is a 7555 IC wired as a monostable multivibrator. The values of C2 and VR2 are such that a short pulse is produced, lasting only a millisecond or so. This

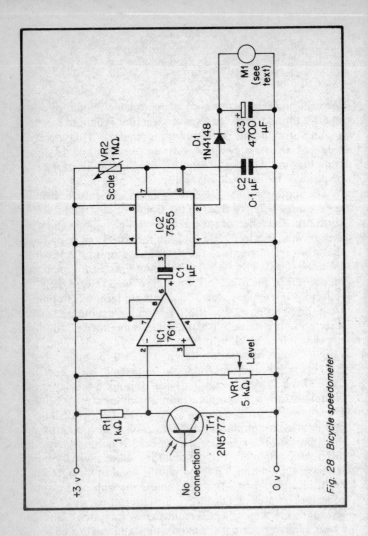

Fig. 28 Bicycle speedometer

brief pulse results in a current flowing through D1 to add charge
to C3. Because C3 has large capacity there is only a slight
increase in the potential of the positive plates of C3.

Each time the bicycle wheel rotates, a pulse of fixed length
and voltage is pumped through D1 into C3. The charge on C3

46

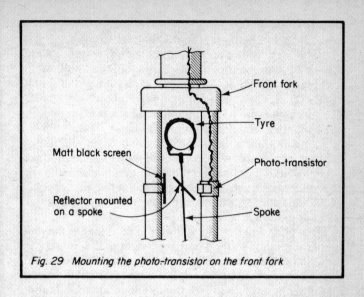

Fig. 29 Mounting the photo-transistor on the front fork

builds up at a rate depending on how frequently the pulses
are generated, i.e. on how fast the bicycle wheel is turning. The
charge slowly leaks away through the coils of meter M1. This
should preferably be a sensitive meter with coils of high
resistance. In the prototype a 30µA fsd meter was used. For
any given rate of rotation of the wheel, there is a balance
between the amount of current pulsing through D1 to the
capacitor and the steady leakage of current through M1. The
faster the bicycle is travelling, the more frequent the pulses
and the greater the charge that is built up on C3. The greater
the charge, the greater the leakage through M1 and the higher
the reading shown by its needle. The position of the needle
thus depends on the speed of the bicycle.

The circuit is best assembled stage by stage. First build
the sensor/amplifier section. VR1 can be a preset potentiometer if required but there is some advantage in using an
ordinary miniature potentiometer with control knob. This
allows the speedometer to be adjusted to work under a range
of different ambient lighting conditions. During testing, the
bicycle may be turned upside-down and the front wheel made
to spin slowly. In a brightly-lit room, enough light should be

reflected on to the sensor to produce the required effect, especially if a white sheet is placed on the floor. The potential at the collector of Tr1 should normally be at about 1.5v to 2v, falling sharply at each rotation of the wheel. When VR1 is correctly adjusted the output from pin 6 of IC1 should be close to zero, with sharp upward pulses at each rotation. The high impedance of the input to IC2 ensures that quite small pulses are sufficient to trigger the timer. This section of the circuit is the next to be constructed. Since VR2 sets the scale factor, which is a once-for-all setting, VR2 is a preset potentiometer. For testing, set VR2 to a fairly high value. This will produce a pulse lasting a second or so. The output of IC2 should be 0v, with a sharp swing to +3v every time the reflector passes the sensor.

If the pulse generated by IC2 were to be as long as the time for one revolution of the wheel, the output of IC2 should stay high. In operation, a very short pulse is required. The length of pulse required depends upon the capacitance of C3, which may differ by as much as 50% from its nominal value. It also depends upon the rating and coil resistance of M1. We are now ready to set the pulse length by practical trials.

First the bicycle wheel is rotated slowly; the needle of M1 flickers at each rotation. Now rotate the wheel as fast as possible, probably the needle will swing firmly to the upper end of the scale. This indicates that the pulses from IC2 are longer than the period of rotation of the wheel. Adjust VR2 to reduce its value, thus shortening the pulses, until the needle settles in a more-or-less steady position near the top of the scale. The next stage is to calibrate the speedometer. This is easier if the reflector is attached to the rear wheel, for then it is easier to crank the wheel at constant speed.

If the tyre has outside diameter 26 inches, its circumference is 81.68 inches. If the wheel were to make 1 turn per minute the bicycle should be moving at a speed of 81.68 inches per minute, which is equal to 0.0773 miles per hour or 0.124 kilometres per hour. The calibration procedure consists in turning the wheel at a known constant speed and marking the dial of the meter accordingly. For slow speeds it is easy to watch the reflector and count the number of rotations per minute. If it makes 65 revolutions per minute the corresponding speed is 65 x 0.0773 = 5 mph. For faster speeds, count the number of rotations of the pedal crank and multiply by the gear ratio. If there are n_1 teeth on the crank wheel and n_2 on the rear axle, the multiplying factor is n_1/n_2. Begin calibration at the greatest speed that is likely to be attained. This depends on the cyclist and the type of machine, but would probably be in the range 15—20 mph. One person should turn the crank steadily to maintain this speed, for example, 259 turns per minute are equivalent to 20 mph. Another person adjusts VR2 to bring the needle of M1 to the top end of the scale. Next a number of slower rates are tried and the position of the needle marked for each rate. It will be noted that the scale of the meter is not linear. Once a few graduations have been marked it is normally easy to mark in the intervening ones by estimation.

Two problems may arise in calibration. The needle may not remain in a steady position even at constant speed. To a certain extent this is inevitable, owing to the nature of the circuit. Increasing the value of C3 will help smooth out excessive jerkiness. If the needle remains near the top end of the scale at slow speeds, this could be because the pulses are too long; try increasing the set value of VR2. It may also be that the current does not leak away fast enough through M1. This can be cured by wiring a resistor in parallel with M1.

The whole unit including the solar cells may be assembled in and on a single case, clamped to the handle-bars. Alternatively, the circuit and cells may be in a case on the rear luggage-carrier, with the meter mounted on the handle-bars.

CHAPTER 7

Timers

All the circuits in this chapter are based on the 7555 timer IC. It has low power consumption and can operate on a supply of 2v or even less. Another of its features is that the length of timing periods is independent of the voltage of the supply. This makes it especially suitable for use with solar cells, provided that the level of illumination remains roughly constant during any long period of use.

Seconds Timer

This extremely simple circuit can be built as a novel vest-pocket timer that has a multitude of uses. Its display is a small and inexpensive microammeter (Fig. 30). In operation the needle flicks from side to side once a second. The circuit can readily be adapted for other timing periods. For example,

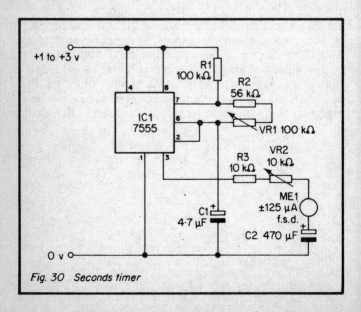

Fig. 30   Seconds timer

it can be built as a 10-second timer, useful in assessing the cost of trunk-calls during peak periods.

The 7555 is wired as an astable multivibrator with a fixed period of 1Hz. VR1 is for setting the fixed frequency and so is a preset potentiometer. Capacitor C1 must be a tantalum bead capacitor. The meter recommended has centre zero. As output from pin 3 rises and falls, current flows into and out of C2 and the needle moves from side to side of its central position. VR2 is a preset potentiometer that controls the amplitude of the motion of the needle.

Although the supply voltage is given as 1v to 3v, most constructors will prefer to use a few solar cells as possible, so will opt for a low voltage. If three cells are used (1.5v supply) R3 may be omitted. If only 2 cells are used there is no need to limit the oscillations of the needle: both R3 and VR2 may be omitted. Even with as few as 2 cells the needle beats clearly and the circuit works well in poor lighting conditions.

Settable Timer

This timer signals when a predetermined period of time has elapsed. It needs a 3v supply but requires only 1.2mA of current, so it can be powered by 6 of the smallest and cheapest solar cells, or by 3 cells and the voltage tripler circuit of Fig. 10. The voltage tripler does not work so efficiently at lower voltages and gives only 2.5v at 1.2mA, but this is sufficient to operate the timer.

The circuit (Fig. 31) makes use of two low power CMOS timers both contained in a single IC the 7556. Timer 1 is connected as a monostable multivibrator, to produce the required time period. Timer 2 is connected as an astable multivibrator to produce the signal that shows when the required period has elapsed. Timer 1 is controlled by two press-buttons of the push-to-make type connected to the "set" (pin 6) and "reset" (pin 4) inputs of the timer. Both inputs are normally held high by the pull-up resistors, R1 and R2. Pressing S1 takes pin 6 low and starts the timing; the output (pin 5), which is normally low goes high and stays high for a period of time determined by the setting of VR1. With the values shown, the timing period can be set for any length up to 10 seconds. If VR1 is replaced with a 1MΩ potentiometer,

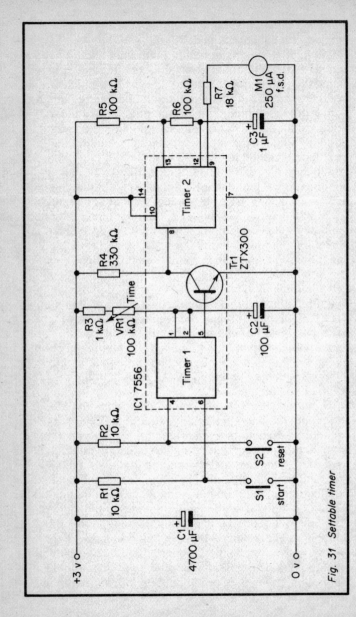

Fig. 31 Settable timer

52

periods up to 100 seconds can be set. This is by no means the limit. Further increase in the length of the period is obtained by increasing the value of C2. Since C2 must be a tantalum bead capacitor and $100\mu F$ is the highest value normally obtainable, two or more such capacitors are wired in parallel to obtain the required capacitance. Resistor R3 is there to prevent excess current flowing to the IC should VR1 be reduced to zero. Increase of timing period may be obtained by increasing R3, but short periods are not possible with this arrangement. For instance, if R3 = 1MΩ and VR1 = 1MΩ, periods between 100 seconds and 200 seconds are obtained.

An alternative method of adjusting timing is shown in Fig. 32. This is useful if a limited number of periods is required, as might be the case for an egg-timer, which could be set to run for 3, 4 or 5 minutes. Whereas VR1 is a standard potentiometer with a pointer knob, the variable resistors in this modification are preset potentiometers.

During the timing period the output of Timer 1 is high, supplying base current to Tr1. This is "on", holding the

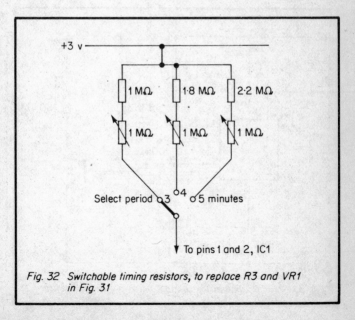

Fig. 32 Switchable timing resistors, to replace R3 and VR1 in Fig. 31

trigger input to Timer 2 (pin 8) low and preventing it from functioning. When the timing period has elapsed, the output of Timer 1 goes low, Tr1 turns off and the trigger input rises. This allows Timer 2 to oscillate at about 2Hz. The pulsing output is fed to M1, causing its needle to flash from side to side. Thus the needle oscillates when the timer if first switched on but is still during the timing period. When this is over the needle begins to oscillate again. At any point during the timing period the timer can be reset by pressing S2, followed by S1.

When constructing this circuit use tantalum bead capacitors for C2 and C3. The other capacitor, C1 is an aluminium electrolytic capacitor, which maintains the potential across the circuit during momentary interruptions of the power supply. The shadow of a passer-by falling on the solar cells will not break the timing sequence.

Build Timer 1 first. Its output may be monitored by a voltmeter connected to pin 5. Check that it operates for the required length of time. Tantalum capacitors have a tolerance

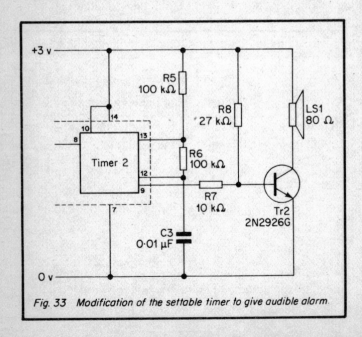

Fig. 33 Modification of the settable timer to give audible alarm.

of 20% so, if the capacitor used for C2 has a lower-than-average value, it may be necessary to substitute another. Next build Timer 2 and test it. The value of R7 may need altering to obtain suitable deflection of the needle of the meter. Finally, connect the timers by way of Tr1.

As an alternative to the visual indicator, the timer can be modified to give audible warning that time is up. As shown in Fig. 33, the frequency of Timer 2 is brought into the audio-frequency range by substituting a capacitor of smaller value for C3. The alternating output from pin 9 is amplified by Tr2 and fed to a loudspeaker.

For both visible and audible signals try the circuit of Fig. 33, which uses a third timer. This is provided by a 7555 IC. The second and third timers may both be triggered by the collector voltage of Tr1.

Metronome

This device is a useful one for musicians and since it is likely to be used for hours at a time, it is economical to run it from solar power. It requires 6 small solar cells to power it. As shown in Fig. 34 the sound comes from a crystal earphone, but the circuit can be modified to give loudspeaker output if desired.

As in previous circuits in this chapter, timing is performed by the 7555 IC. During operation, C1 and C2 are charged through R1, R2 and VR1, but are discharged through R2 and VR1 only. Discharging is more rapid than charging. The consequence of this, as shown on Fig. 35, is that the output from the timer does not have a 1:1 mark-space ratio. Such a ratio is essential in a musical application in order to provide a regular beat. To obtain an exact 1:1 mark-space ratio we feed the output from the timer to a J–K flip-flop (Fig. 36). The J and K inputs are both held high and the effect of this is that the output Q changes state when the clock input changes from low to high. This is shown in the lower curve of Fig. 35. The output of the flip-flop has an exact 1:1 mark-space ratio and has half the frequency of the clock. The output of the flip-flop is amplified by Tr1. At each change of state a sharp click is heard from the earphone. This click is distinct and in a quiet environment it is not necessary to wear the earphone in order

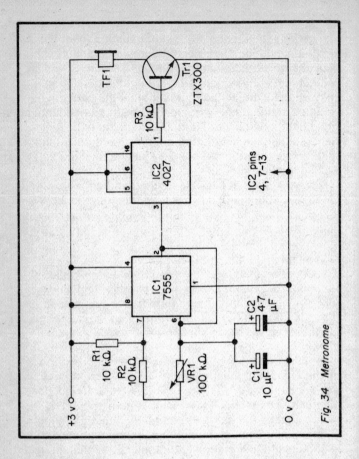

Fig. 34 Metronome

to hear the beat.

Construction presents no problems. It is important that C1 and C2 are tantalum bead capacitors. VR1 should be a standard carbon potentiometer with linear track. IC2 contained two flip-flops, only one of which is used. The inputs to the unused flip-flop must be grounded and are listed on Fig. 34. When the circuit is assembled and power is applied, loud clicks should be heard from the earphone. The rate of clicking depends on the setting of VR1. With the component values given, the metronome covers the range 25 to 200 beats per minute. Even

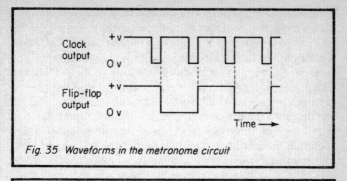

Fig. 35 Waveforms in the metronome circuit

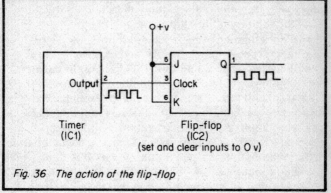

Fig. 36 The action of the flip-flop

allowing for variation in the actual values of components, this allows for the knob of VR1 to be calibrated for all the musical *tempi*.

Stopwatch

The problem of obtaining a good visual display with low powered equipment was discussed on p.40. Another solution is the use of liquid crystal displays. These operate with a few microwatts of power and, like the microammeter display we have already described, rely on ambient illumination to provide the power by which we actually see the device. Unfortunately the manufacturers have not provided constructors with the equivalent of a single lamp – a small LCD panel that goes light or dark when signal is applied to it. LCD

57

displays usually consist of 4 or more 7-segment digits together with other symbols. This makes them expensive and also limits their use to those projects that require numeric displays. A solar-powered clock is feasible, but would need battery back-up to cover periods of darkness. Many of the published designs can be adapted to solar power simply by adding the appropriate number of solar cells and nickel-cadmium cells together with the constant current IC. Chapter 3 provides information on this. The device described here is intended for operation for only limited periods during which it may be assumed that the level of light will stay reasonably constant. The 7555 timer IC is not affected by moderate changes in the supply voltage, so minor fluctuations in light levels will not effect the accuracy of the stop-clock.

The block diagram of Fig. 37 shows how the stop-clock works. The timer produces pulses at 100Hz and these are fed to a train of divide-by-ten counters (4518). Each of the counters has a 4-bit binary output that runs through the sequence 0000, 0001, 0010, 0011, ..., 1001 and back to 0000, on the arrival of successive pulses. The output from the counters goes to four decoders (4055). These convert the 4-bit BCD input to the seven outputs needed to produce the corresponding decimal digits on the 7-segment liquid crystal display. The display runs through the sequence from 00:00 to 99:99 and then repeats.

The circuit is an economical one in that the same timer IC is used to provide the display frequency. Its 100Hz output is suitable for providing the alternating field needed by the LCD. This "display-frequency" is fed to the 4055 ICs and to the back plane of the display. The action of the 4055 ICs is two-fold. For those segments that are to be displayed *black* the square-wave sent to the segment electrode is *out of phase* with the square-wave sent to the back-plane. Thus when the back plane is at +5v the segment electrode is at 0v; when the back plane is at 0v, the segment electrode is at +5v. Consequently, there is always a pd of 5v between the segment electrode and the back plane; the molecules of the liquid crystal become aligned *along* the field and the segment appears black. For those segments that are not to be displayed, the square-waves are *in phase*. There is always zero pd between the electrode and back plane; the molecules become helically arranged *across*

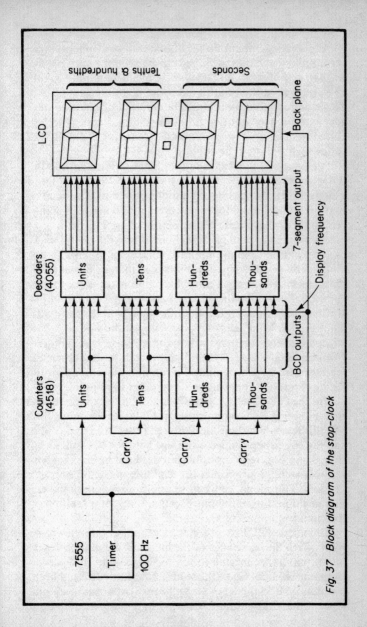

Fig. 37 Block diagram of the stop-clock

the field and the segment appears clear.

Although this circuit uses 7 ICs and has a prominent display, it consumes only 5mA at 5v and can be powered by 10 of the smallest size of solar cells. However, a reduction in the number of cells required can be achieved by using the voltage tripler. This requires only 2.7v to drive the stop-clock, so 6 cells are adequate, even in light that is less than full sunlight. The current requirement with the voltage tripler is still only 25mA, so the solar cells need be only small ones.

Fig. 38 shows the details of the circuit. VR1 is a preset potentiometer to allow the frequency of the 7555 to be set to exactly 100Hz. It is best to build this section of the circuit first, since the display frequency is required when testing the remainder of the circuit. Since there is a large number of connections to be made in this circuit it is worth considering the possibility of etching a printed circuit board. This is particularly worth doing if it is intended to build a pocket-sized clock. Take special care when drawing (or wiring) the connections between the decoders and the socket of the LCD. It will be seen from Fig. 38 that the colon between the right-hand and left-hand pairs of digits is brought into use in order to separate seconds from fractions of seconds. The colon is permanently displayed by supplying it with the display frequency *inverted* by Tr1. This produces a continuously alternating field between the colon electrodes and the back plane.

The lead between IC1 and IC2 is liable to pick up interference that induces IC2 to count, even when it is receiving no input. For this reason a screened lead is used between S2 and IC2, and the screen is grounded. The lead between IC1 and S2 should also be screened, although it need not be if it is only a few millimeters long. Timer ICs tend to generate excessive "noise" so C2 should be wired so as to absorb voltage spikes that might pass from the timer section to the other section of the circuit.

The input of IC2 goes to the *"clock"* input (pin 9), which responds to the *rising* edge of the incoming pulse. The "enable" input (pin 2) is connected to +5v to allow counting to occur. In the succeeding stages (the second counter of IC2 and both counters of IC3) the input is the "carry" from the fourth bit (Q4) of the previous stage. At these stages, a count is to occur when this output goes *low*. The "carry" is taken to the *"enable"* input (pin 2) instead of to the "clock" input, since the enable input responds to a *falling* edge. The "clock" input of these stages is grounded. The reset inputs of all counters are held low by the pull-down resistor R5. To reset the counters, button S1 is pressed, making the reset inputs temporarily high. Note that S1 is a *push-to-make* button, but S2 is a *push-to-break* button.

Since CMOS ICs do not work reliably with floating inputs, it is best to complete the wiring of the whole circuit before testing. When all is complete and checked, switch on the power and observe the display. This should show clear black figures. If it is blank, including the colon, investigate the display-frequency circuits. When the clock is first switched on, the display should be showing the rapid counting action in progress. The two right-hand figures will be changing too rapidly for the individual numbers to be distinguishable. If S1 is pressed, the display should immediately change to "00:00". If it does not, investigate the reset circuit. There are so many soldered joints in the device that it is almost inevitable that there will be one or two "dry" ones. Re-solder any that are of doubtful status.

As soon as S1 is released, counting begins again, from "00:00" up to "99:99". Counting stops whenever S2 is pressed. If it does not, check the earthing of the screened lead. Stop the counting several times to check that all figures in the display are working properly. If the symbols "L", "H", "P", "A", "—" or a blank appear, there is a mistake in the connections between the counters and the decoders. The decoders produce these symbols when fed with counts between 1010 and 1111. In this circuit this can happen only if wires are crossed. If the display produces strange symbols, neither numerals nor letters, then the wiring between the decoders and the display is faulty. Occasionally one of the spring contacts in the socket may be bent out of position.

The timer may be calibrated by comparing it with a watch or another stop-clock. Time it for a fairly lengthy period — say 100 seconds or more. Adjust VR1 carefully until it runs accurately. To use the stop-clock for timing events, first press the reset button and hold it pressed. Then, to begin timing,

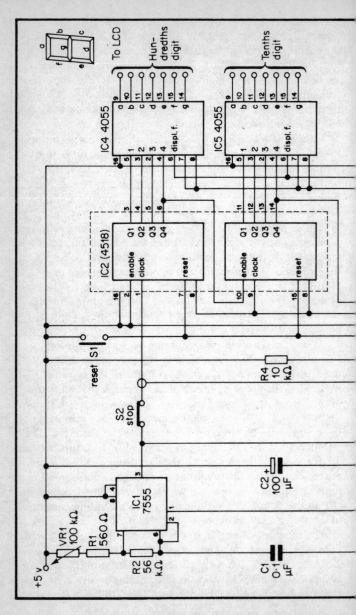

62

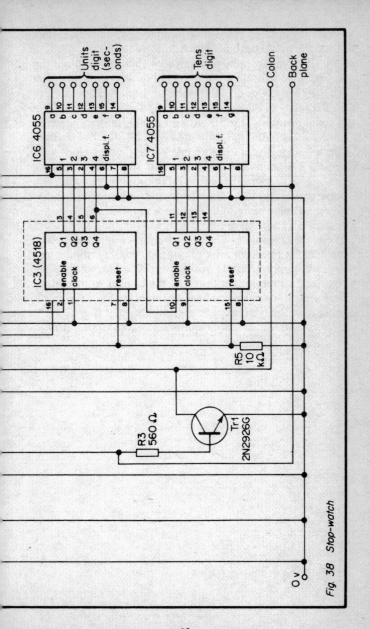

Fig. 38 Stop-watch

63

release the button. When time is up, press and hold the stop button and read the time displayed.

The design presented here can be used as the basis of other timers. Two separate timer ICs could be used (or a dual timer IC, the 7556). One could provide the display frequency (50–200Hz) and the other could operate at a much lower frequency. It could operate at 1.67Hz, giving a display of minutes, with tenths and hundreth of a minute. It could operate even more slowly, producing one pulse a minute, so the stop clock would run for 9999 minutes (almost a week). Another possibility is to use only one timer IC but to interpose another counter between IC1 and IC2. This divides the frequency by 10, giving a display of up to 999.9 seconds in tenths.

This is a relatively simple circuit and there is a small source of error in the timing. The count begins when S1 is released but the first pulse generally occurs *sooner* 0.01 second later, depending upon the state of the timer IC at the instant of release. This means that the timer can be in error by up to 0.01 second. Such an error can be neglected for errors from other causes may possibly be greater. Another timer, entirely different in design and function appears on p.100.

CHAPTER 8

Measuring Temperature

These two projects measure temperature in two different ways. There are many ways in which these projects may be modified to perform various functions.

Temperature Alarm

This sounds an alarm or activates a visual indicator whenever the temperature falls below or rises above a preset value. It operates from 6 small solar cells. Fig. 39 shows the circuit that is triggered when the temperature exceeds the preset value. It can be placed in a greenhouse to sound an alarm when the greenhouse becomes too hot. The circuit and solar cells will be in the greenhouse, with the loud-speaker or indicator indoors. The circuit operates only during the day but this covers the whole of the period of risk. The sensor R1 is a thermistor with negative temperature coefficient. That is to say, its resistance decreases as temperature increases. The thermistor and resistor VR1 act as a potential divider. The potential at point A depends upon the ratio of the two resistances. In use, VR1 is set so that the potential at A is just below about 1.5v. This acts as a "low" input to the first of the NOR gates. All three gates are wired with their two inputs joined together, so each acts as a NOT or INVERT gate. Consequently, the signal from A is inverted 3 times and the output of the third gate is "high". The output is fed to an alarm that sounds on intermittent note. This is basically the same as that shown in Fig. 24, but here it is controlled by the input to pin 13, instead of by a switch. The way the alarm generator works is described on p.38. Just as the output from the slow oscillator inhibits the fast oscillator, so the output of the third inverter of IC1 inhibits the slow oscillator.

At $25°C$ the resistance of R1 is approximately $47k\Omega$. As temperature rises the resistance of R1 decreases. Consequently, the potential at A rises. With sufficient rise in temperature, the potential at A acts as a high input to the first gate. Its output goes "low", the output of the second gate

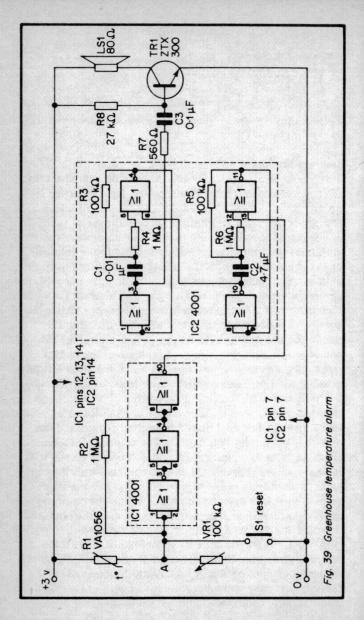

Fig. 39 Greenhouse temperature alarm

66

goes "high" and the output of the third gate goes "low". The "low" input to pin 13 of IC2 causes the alarm to sound.

The resistor R2 acts as a feedback resistor. As the potential at A rises *slightly* above a given level, the gates *begin* to change state. The rising output from the second gate is fed back through R2 to the input of the first gate, causing a further rise. This *positive feedback* ensures that any slight excess of potential at A results in a firm triggering action. Moreover, once triggering has occurred, the potential at A is pulled a little higher than that determined by the ratio of R1 to VR1. This means that the temperature must fall some way *below* the level at which triggering occurred before the circuit reverts to its previous state and the alarm is switched off. This then means that the alarm does not cease should temperature hover around the "turn-on" point. Push-button S1 allows the circuit to be reset. Although S1 is optional, it is useful to have this facility, particularly when adjusting VR1 to the position required.

It is better to build the alarm section of the circuit first, as described on p.39. If the device is to be sited in the greenhouse, the loudspeaker will be on a long lead and be mounted indoors. The alarm is tested by connecting pin 13 to 0v, when the alarm should sound. Connecting pin 13 to +3v should silence the alarm. Next construct the trigger section of the circuit. The type of thermistor recommended is an inexpensive rod thermistor. It should be mounted in a central location in the greenhouse, but not where it will receive direct sunlight. It could be mounted in a small box, with perforated walls. Another point to watch is that the thermistor must not be placed where it can become wet when the plants are watered.

The same circuit can operate in the reverse direction, to trigger the alarm when the temperature falls below a certain level. As such, it has uses as a frost detector. Normally it would be set to trigger *before* freezing-point is reached, say at $1°$ or $2°C$. Even the most enthusiastic gardener will not relish being woken up during the night, so it is unnecessary to provide supplementary battery power. If the main circuit is housed indoors, with the thermistors in an exposed situation outdoors, normal room lighting can be used to provide power for the solar cells. The only modification to the circuit is to interchange R1 and VR1. As temperature falls, the resistance of R1 increases, causing the potential at A to rise and trigger the alarm.

Desk Thermometer

The desk thermometer has few components. It operates on only 1.5v so only 3 small solar cells are needed. The unit can be constructed in a very small case making it ideal for desk-top use. This is not the only way in which the circuit can be applied. The characteristics of its temperature sensor mean that the sensor can be mounted at some distance from the rest of the circuit. It can be used to monitor the temperature in an incubator, deep-freezer, greenhouse or loft, the temperature reading being shown on a meter located in the living room or work-room. By day, the instrument can receive light from a nearby window. By night it can be placed beneath the desk-lamp. The circuit (Fig. 40) is based on the constant current IC that we have previously used in battery-chargers (Chapter 3). Current is independent of applied voltage over a wide range, so the reading is not affected by fluctuations in illumination received by the solar cells. Since the IC is a constant current device, the current leaving it is set by R1 and the whole of that current must flow from point A to point B no matter how long the wire between A and B. This is why the sensor can be mounted in a remote and possibly inaccessible position if desired.

The current from IC1 is determined not only by the value of R1, as explained on p.19, but also by the ambient temperature. The current flowing into V+ is

$$I = (227 \times T)/R$$

where I is in microamperes, R is the value of R1 in ohms and T is the temperature in kelvin. On the Kelvin scale, the boiling point of water is $373°K$, the melting point of ice is $273°K$ and absolute zero is $0°K$. The temperature difference of a degree on the Kelvin scale equals that of a degree on the Celsius scale (or Centigrade scale). Thus if $R1 = 100\Omega$ and the temperature is $0°C$ ($= 273°K$), the current is

$$I = (227 \times 273)/100\mu A = 620\mu A$$

The current flows on through R2, causing a pd to appear across

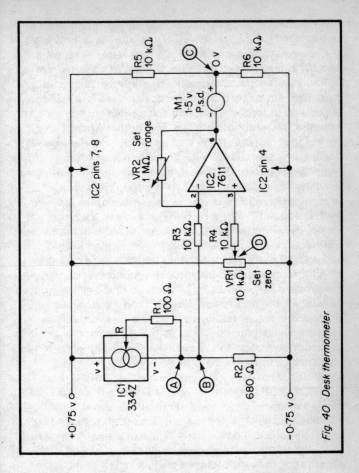

Fig. 40 Desk thermometer

R2. This is measured by the operational amplifier IC2, connected as a differential amplifier. Ignoring the small fraction of the current passing along R3, toward the "virtual earth" at pin 2 of IC2, the current flows along R2, causing a potential of 620 x 680 = 421600mV or 0.42v to appear at point B. This value is relative to the −0.75v line, so it is −0.33v with respect to the 0v line of the operational amplifier (point C). If VR1 is adjusted so that the potential at its wiper (point D) is *also* −0.33v, the differential between the inputs of the amplifier is zero. Its output is zero and this is indicated by the meter M1. If temperature increases *above* 0°C, current is increased. At 20°C, for example, it becomes 665μA. This *rise* in voltage causes the output voltage to *fall*, but by a greater amount, since voltage differentials are being amplified. The meter is connected so that a *positive* reading appears on its scale as the output voltage falls. In this example, the setting of VR1 gave zero output when temperature was 0°C. Other settings of VR1 can give zero reading at other voltages. VR1 is used to decide which temperature is to be the lowest that the instrument can indicate. The change in output voltage for a given change in input differential is decided by the setting of the feed-back resistor VR2. The greater the set value of VR2, the greater the amplification. This resistor is used to set the *scale* of the instrument. For example, we can use VR1 to determine that the lowest point on the scale is 0°C and then set VR2 to make full-scale deflection equivalent to a temperature rise of 50 degrees. The scale of the meter can then be graduated to read from 0°C to 50°C. If we are more interested in precise readings over a more limited range, we can set the minimum reading to 20°C and make full-scale deflection equivalent to a rise of 10 degrees. The meter is then graduated to read from 20°C to 30°C.

Since IC1 has only 3 terminals and is the size of a small transistor and IC2 is only an 8-pin DIL device this circuit can be built on a very small board. Possibly the circuit could be accommodated within the casing of the meter, with the two solar cells mounted on the back. The sensor must be mounted so that the air of the room can circulate freely around it. If the sensor is to be mounted remotely, R1 must be mounted as close to it as possible so that both experience the same temperature. The accuracy of the instrument depends upon the temperature stability of R1, which should preferably be of the thick-film metal-glaze variety. The terminals of the sensor IC and R1 must be soldered. If a socket were to be used for IC1, changes in contact resistance at the socket could seriously upset the accuracy of the instrument.

CHAPTER 9

Touch Alarm

A typical application for this project is for use by an elderly or infirm person when signalling for help. If it is to be used only by day or under artificial illumination, it can be powered directly by 6 small solar cells, or by 3 cells and the voltage tripler (p.17). A rechargable nickel-cadmium battery is needed if it is to be used in darkness. Alternatively, it can be powered by solar cells except during the night, when a dry battery is switched in to replace the solar cells.

The complete circuit is shown in Fig. 41. Part of this (IC2) is recognisable as the intermittent alarm generator, although this version employs NAND gates instead of NOR gates. The circuit works by sensing the alternating voltage induced in the human body when it is in the magnetic field surrounding mains cables and mains-powered equipment. The currents that such voltages produce are extremely minute but the voltage changes are readily detectable by a field effect transistor (Tr1). The gate of this has very high input impedance so it can follow the voltage changes of the body without drawing appreciable current.

When a finger is placed against the touch-plate, an alternating and amplified potential develops at the drain terminal of the transistor (point A). Rising potentials cause a current to flow to the base of Tr2 to turn that transistor on. The voltage at its collector begins to drop. The falling voltage is applied to one of the NOR gates of IC1 which acts as an inverting amplifier, turning a slowly falling voltage into a rapidly rising one. In other words, it generates a "high" pulse on the line from pin 3 to pin 12. Before the plate was touched, the voltage level on the 3−12 line was low. Now a high pulse triggers the flip-flop to change state. Previously the output of the flip-flop (pin 10) was low and the alarm was silent. When the plate is touched, the output of the flip-flop goes low and the alarm begins to sound its intermittent note. It will be noted that, because the alarm circuit is based on NAND gates, it is inhibited by a low input to pin 1 and activated by a high input, the reverse

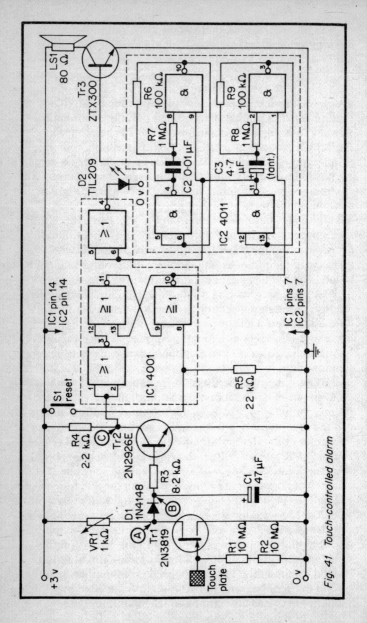

Fig. 41 Touch-controlled alarm

response to that of an alarm based on NOR gates. Since pin 1 is low in the inactive state, output at pin 3 is high. Output at pin 11 is the INVERT of this; it is low, so inactivating the fast multivibrator. The output of pin 11 is also fed to the fourth gate of IC1 and is inverted by it, turning on a LED (D2). This lamp acts as an indicator that the power is switched on. If the circuit is to be operated under high-light conditions in which a LED could not show clearly, a cheap microammeter could be substituted for D2. Just as the output from pin 11 is low when the device is quiescent, so is the output from pin 4. This means that Tr3 is off and no current is passing through the loudspeaker. This design keeps the current requirements to a minimum when the alarm is not being sounded. As mentioned above, D2 is normally lit to indicate that the circuit is ready for action. When the alarm is sounding the slowly alternating output of the slow multivibrator causes D2 to flash at about 1Hz, indicating that the circuit has responded to the touch. The alarm has only a moderate volume; it would be inappropriate for a simple request for a cup of tea to rouse the whole neighbourhood! But if placed at the bedside it can certainly awaken all but the most heavy sleeper.

Once the alarm has been triggered, it sounds until button S1 is pressed. This sends a high pulse to the other input of the flip-flop, causing it to revert to its original state. The function of VR1 is to vary the potential at point A so that the small increase due to touching the plate is sufficient to trigger the circuit. This is a once-for-all adjustment, and VR1 can be a preset potentiometer. Fig. 41 shows the 0v rail connected to Earth. This *may* be necessary on some circumstances. It is best to try the circuit without the Earth connection, for it will generally work without it.

The prototype was built so as to be as compact as possible and for operation at night. The final design is shown in Fig. 42. The circuit was assembled on a small fragment of strip-board cut to fit inside the plastic cap of an aerosol insecticide spray can. The touch-plate was mounted on top of the cap. The plate was a burnished steel plate manufactured for this purpose (Maplin Electronics, p.110) and has a 6BA threaded rod. A solder tag was attached to this. The material of the cap was a white translucent plastic. It was drilled very carefully to take

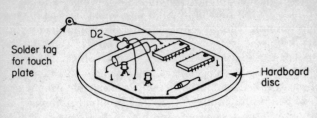

(a) Circuit board mounted on hardboard disc

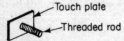

(b) Underside of touch plate

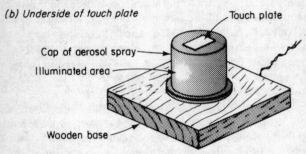

(c) Completed touch-switch

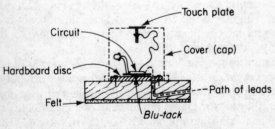

(d) Section through the block

Fig. 42 One way of building the touch-controlled alarm

the rod of the touch-plate, as this material is rather brittle.

Construction of the circuit itself presents few problems, except possibly those of working on a compactly laid out board. Some readers may prefer to design and etch a pcb to obviate wiring difficulties. The loudspeaker is to be housed in a separate enclosure. If the design of Fig. 42 is being followed, the loudspeaker cabinet could also contain the power supply (a solar battery, with or without voltage tripler, or a dry battery). The reset button is also best mounted on the speaker enclosure. It is preferable to begin construction with the alarm generator and test this thoroughly. Then build the flip-flop and wire it to the alarm generator. Both inputs to the flip-flop should be normally low; to test its action, temporarily connect pin 12 to the 0v rail. Pin 8 is already grounded by way of R5. Connecting pin 12 to high makes the output at pin 10 go high (+3v), while connecting pin 8 to high (with pin 12 low) makes the output at pin 10 go low. The final stage to be built is the triggering circuit. Measure the voltage at point B when testing this part of the circuit. Adjust VR1 to bring the voltage at B to about 0.6v. A light touch on the plate should then cause the voltage to rise by a few tenths of a volt. The voltage at point C may then be tested to see if a correct response is obtained. It should be above 1.5v but fall well below 1.5v when the plate is touched.

In the prototype, D2 was mounted so that it pointed towards one side of the cap. When the LED is on, a circle of light is seen from the outside, sufficient to locate the switch in total darkness. When the plate is touched the flashing of the LED is clearly seen. To fix the plastic cover in place the following method was used. First two holes were bored through the wooden base to carry the leads to the speaker and power supply (Fig. 42(d)). A circular disc of hardboard was cut to be an exact fit into the cap; most caps have small ridges that ensure that they clip firmly on to a circular can-top and, with care in shaping the disc, a firm grip can be obtained. The disc was nailed to the base board, the hole in the disc coming above the vertical hole in the base. The circuit board was fixed to the disc by a pellet of *Blu-Tack*. When all has been tested. and found to be in order, a rectangle of felt was stuck to the bottom of the base to protect the table-top on which it was to stand.

CHAPTER 10

Some Audio Projects

Morse Code Practice Set

This circuit can be used by a person for practicing the Morse code, or for sending messages from one station to another. Fig. 43 shows the simple practice set, that can also be used

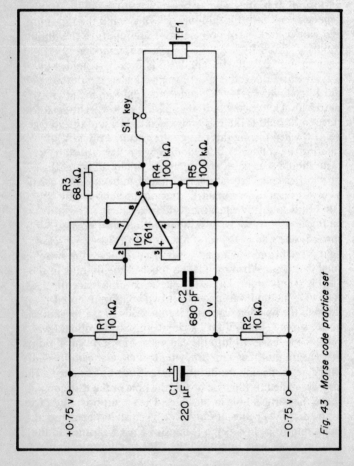

Fig. 43 Morse code practice set

for one-way signalling. Fig. 44 shows the wiring for a two-way system. Both systems have very small power requirements, and operate at low voltage, so only 3 of the smallest and cheapest solar cells are required.

The oscillator that produces the tone for this circuit is based on an operational amplifier. Its output (pin 6) is divided by the potential-divider R4—R5 so that half of it is fed back to the non-inverting input. This is positive feedback, leading to instability and, consequently, to oscillations. The period of oscillation is determined by the values of the resistor R3 and capacitor C2. When output goes high, input is high at pin 3. Current flows from pin 6, through R3 and gradually charges C2. As the potential at pin 2 reaches the level of that at pin 3 the output begins to fall. It goes negative and now C2 becomes gradually discharged through R3 to the negative potential at pin 6. With the values of R3 and C2 given, a high-pitched note of a few thousand hertz is heard when S1 is pressed. In this application S1 is the morse key.

The circuit takes up little space. It could be contained in a flat rectangular box with the two solar cells and the morse key mounted on top. TF1 is a crystal earphone as used with an ordinary portable radio set. This could be replaced by an 80Ω loudspeaker if a higher operating voltage, say ± 1.5v, was employed. Only one oscillator is required for two-way transmission as it is usual for messages to be passed only one way at a time. Provided that the users are familiar with the "over to you" routine, only two wires are required to join the stations (Fig. 44). The additional switches must be set to "send" or "receive" as required. When messages are not being sent, one station is to be left on "receive" and the other on "send". Only the "send" station can then call the attention of other. To provide a system in which either station can call the other we need three connecting wires, as in Fig. 45. This is more expensive, particularly if the distance between stations is great, but has the advantage that the "send-receive" switches are not required. Details of the Morse Code are shown in Fig. 46.

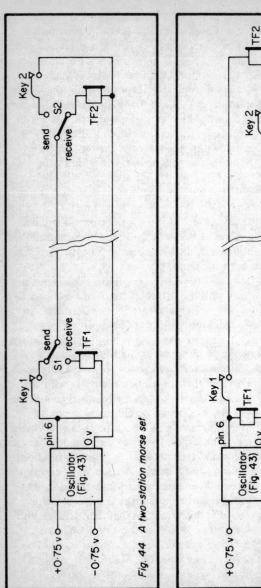

Fig. 44 A two-station morse set

Fig. 45 A two-station morse set which allows either station to call

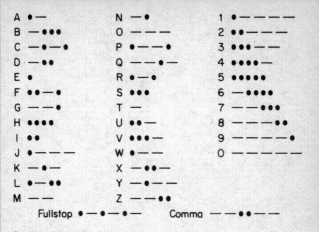

A •—	N —•	1 •————
B —•••	O ———	2 ••———
C —•—•	P •——•	3 •••——
D —••	Q ——•—	4 ••••—
E •	R •—•	5 •••••
F ••—•	S •••	6 —••••
G ——•	T —	7 ——•••
H ••••	U ••—	8 ———••
I ••	V •••—	9 ————•
J •———	W •——	0 —————
K —•—	X —••—	
L •—••	Y —•——	
M ——	Z ——••	

Fullstop •—•—•—   Comma ——••——

Consider:

A dot to be equal to one unit.

A dash to be equal to three units (i.e. three times as long as a dot).

The space between individual characters of a letter to be one unit.

The space between letters to be three units.

The space between words to be seven units.

Fig. 46 Morse code

Intercom Systems

The circuit for the simplest type of intercom is shown in Fig. 47. It allows one-way communication. Power requirements are extremely low, for only ± 1.3v is needed to provide an audible signal. At this voltage the circuit requires only $120\mu A$ so that 3 of the smallest and cheapest solar cells are enough to power the circuit even under relatively dull conditions. However, if an extra cell or two is added, volume is appreciably increased. Very good transmission is obtained by using 6 solar cells together with the voltage tripler. The combined current requirements for the tripler (Fig. 10) and the intercom are still only 15mA, so cells of small area will be more than adequate. The reader might experiment to extend the diode-pumping section of the tripler to obtain greater

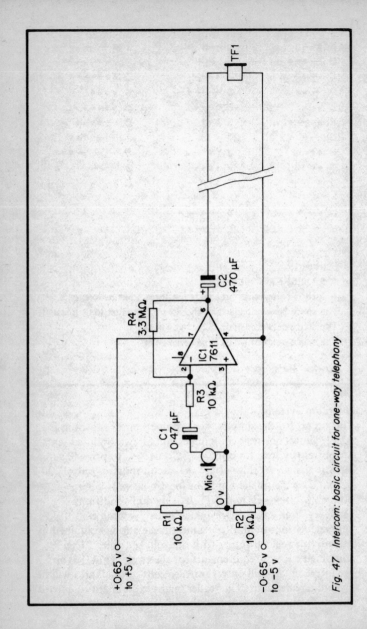

Fig. 47 Intercom: basic circuit for one-way telephony

voltage magnification and thus operate the intercom at high volume with, say, only 3 solar cells.

The intercom relies on the low operating voltage of the 7611 operational amplifier. Apart from that feature, the circuit is a conventional inverting amplifier with nominal gain of 330. Construction is simple and needs no comment except to point out that pin 8 of IC1 is left unconnected. The microphone is a cheap crystal microphone insert or a microphone may be "borrowed" from a tape-recorder. Similarly the earphone is a cheap crystal earphone of the kind used with portable radio sets and tape-recorders.

A signalling facility is generally needed to call a person at the other end of the line. An entirely separate system could be used, such as the buzzer or intermittent alarm of Fig. 22 or 24. To use this we would need an extra wire between stations, the 0v line being shared by both systems. A simpler approach is to use the op amp as an oscillator, when it is not being used as an amplifier. Fig. 48 shows how the conversion may be affected. A three-pole two way rotary switch is used to change the circuit around the IC from an audio amplifier to an oscillator. The oscillator operates as described on p.77. When the system is not in use, S2 is switched to "stand-by" and S1 is switched to "Voice". To speak to a person at the other end of the line the caller switches S1 to "Calling signal". The oscillator then generates a low-frequency signal that causes the needle of M1 to vibrate vigorously. This is intended to catch the attention of the called person. Ways of making the meter more conspicuous are described on p.41. For this circuit a centre-zero meter should be used. On seeing the signal the called person switches S2 to "receive" and waits for the caller to switch S1 to "voice" and speak the message.

Similar systems can be built using a loudspeaker instead of an earphone. This gives much greater volume, but it requires a greater minimum voltage and a higher current to operate it. The minimum voltage is ± 1.5v, so 6 solar cells are required. The current is 1.5mA, so only small cells are needed. This circuit is not suitable for operation from the voltage tripler when a loudspeaker is used. As a simple one-way intercom the circuit of Fig. 47 is modified as in Fig. 49. Since additional amplification is provided by Tr1 the amplification of IC1 must

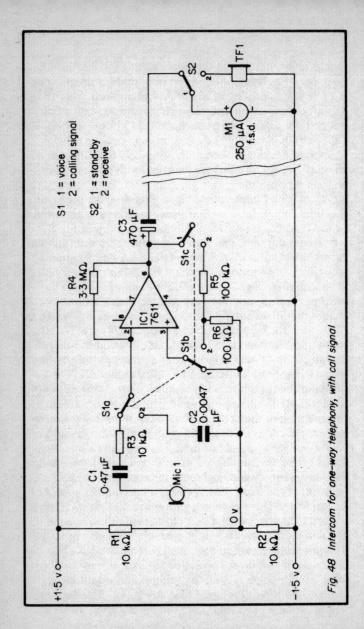

S1 1 = voice
    2 = calling signal

S2 1 = stand-by
    2 = receive

Fig. 48 Intercom for one-way telephony, with call signal

82

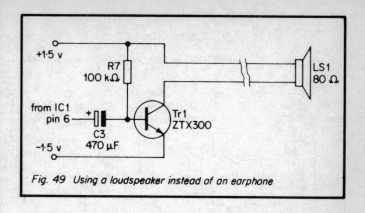

Fig. 49 Using a loudspeaker instead of an earphone

be slightly reduced. The feedback resistor R4 is replaced by one of 1MΩ. This could be a 1MΩ preset potentiometer, allowing control over the volume.

The loudspeaker can be used to call persons to listen by sending an audio-frequency signal to it. The IC is used to generate this frequency. The circuit is as in Fig. 48, except that C2 has the value 180μF, R4 has the value 1MΩ and the whole of the right-hand portion (S2, M1, TF1) is replaced by Fig. 49.

Fig. 50 shows how two identical stations, very similar to that of Figs. 48 and 49 except for the switches, can be linked to make a two-way intercom. The stations share common power supply from 3 medium-sized solar cells. If the stations were to be powered independently, each with its own battery of solar cells or dry cells, the +1.5v line that joins the stations in Fig. 50 could be omitted.

Control of each station is effected by a 3-way 4-pole rotary switch. When the station is left on stand-by the switches are in position 2. Station "A" is in this position in Fig. 50. This is also the "receive" position. To call the other station, S1 is turned to position 1 which converts the amplifier into an audio-frequency oscillator and sends a signal to the loud-speaker of the other station. During conversations both operators turn their switches to positions 2 or 3, as required.

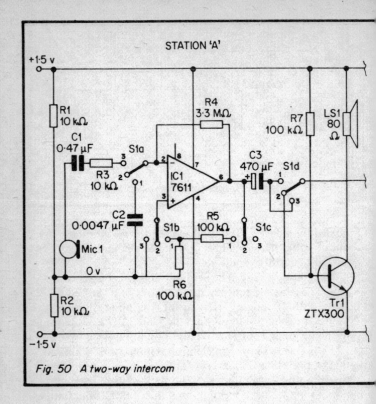

Fig. 50   A two-way intercom

Personal Mini-Organ

This novelty circuit produces a pleasantly musical sound and,
since the sound comes from an earphone, the family will not
be distracted when it is being played. The more proficient
musicians may adapt the design to provide loudspeaker output
(see Fig. 49).

The heart of the organ is the 4046 CMOS IC which is
described in the catalogues as a phase locked loop IC. Here we
are using *part* of the PLL circuitry contained in this IC — the
voltage-controlled oscillator. The basic frequency of the
VCO is determined by the values of R1, C2 and the combined
resistance of VR1 and R2. VR1 can be adjusted to give any
preferred range of notes. Once VR1 has been set, the frequency

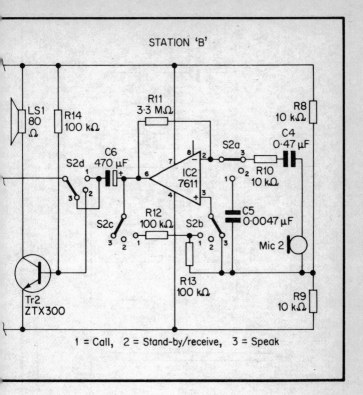

1 = Call,   2 = Stand-by/receive,   3 = Speak

of the oscillator is controlled by the voltage applied to pin 9.
Fig. 51 shows a variable resistor connected as a potential
divider, so that any voltage between 0v and +3v can be
applied to pin 9. This potential divider can take many forms
and the exact form it takes has a great effect on the kind of
music the instrument makes.

The simplest "electronic" approach is to wire a potentiometer (about 10kΩ) between the 0v and 3v rails and connect
pin 9 to the wiper. Tunes are played by rapidly altering the
position of the knob. To make the circuit more like a true
musical instrument use a length of high-resistance wire held
straight on a board. It is connected between 0v and +3v with
a flying lead connected to a probe (banana plug). This probe

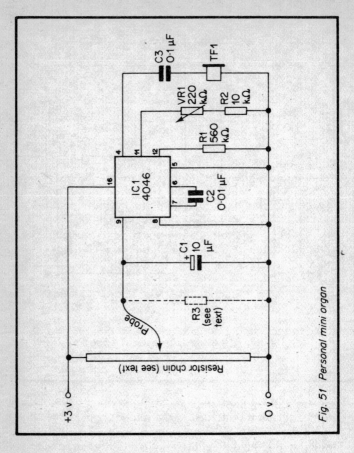

Fig. 51 Personal mini organ

is touched against the wire at different places to produce the different notes required. The board can be marked to indicate the positon for each note. The probe can be jumped from note to note, or made to slide, so as to produce a variety of musical effects.

Fig. 52 shows another way of providing a variable voltage. The two touch plates can be strips of copper mounted with a 1mm gap between them, or adjacent strips on a piece of stripboard. The strips are bridged by one or more fingers *(slightly moistened)*. By altering the amount of pressure, as well as the

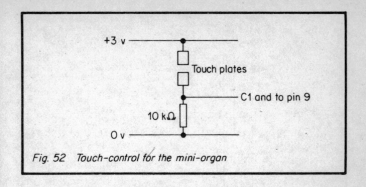

Fig. 52   Touch-control for the mini-organ

number of fingers used, a varying voltage may be produced.

A truly "solar" method is to use a solar cell to produce the voltage (Fig. 53). By using the hands to partly shade the cell, a varying voltage is obtained. The instrument can be played without actually touching it by hand.

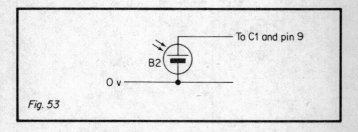

Fig. 53

If a true musical scale is required it is possible to construct a resistor chain that provides a set of voltages. Each voltage corresponds to one note on the scale. Fig. 54 shows how a single octave may be covered by a number of fixed resistors and variable resistors wired in series. The terminals could be a row of metal contacts (e.g. drawing pins) on a base-board and a probe touched against them one at a time. Each variable resistor is adjusted to tune each note to the true scale. A more expensive arrangement is to wire 8 switches or push-buttons to connect the 8 tapping points to C1 and pin 9. The switches are to be closed one at a time. The resistor chain of Fig. 54 has a large resistor at its top end. This could be

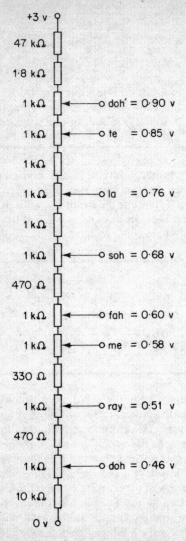

Fig. 54 Resistor chain for one octave.

replaced by a second chain of smaller resistors (fixed and variable) to give notes several octaves higher. Owing to variations between resistors it is not easy to state exactly what values will be required. If the chain is to be extended, some experimentation is necessary.

Pin 9 is a high-impedance input so very little current is drawn when it is connected to potential-divider networks. Another consequence is that when the probe is removed from contact with the network the pitch of the note remains unchanged for several tens of seconds. After the probe has been touched to a point at *different* potential, a fraction of a second elapses before C1 becomes charged to the new potential. This gives a slight *glissando* (sliding note) effect between one note and the next. The amount of this effect can be increased by increasing the value of C1. Two or three capacitors of different value could be wired so that any one can be switched into circuit. This would allow different amounts of *glissando* to be selected. Another effect that is interesting to experiment with is to wire a high-value resistor at R3. This produces an immediate drop in pitch as the probe is removed from the contact. This is a simple circuit but it gives plenty of scope for the inventive reader to produce a wide range of musical (?) effects.

CHAPTER 11

Photographic Exposure Meter

This is by far the simplest project in the book but is is a very useful one. Even though you have a built-in exposure meter in your camera, there are many occasions when a separate meter such as this can help you obtain precisely the exposure effect that your require. For example, if a particular part of the scene is of special interest you can measure the appropriate exposure for that area alone, even though the camera takes in more than that area. A separate meter is also required when determing exposures by the incident light method, a method that many consider to be superior to the reflected light method necessarily used with built-in meters. In addition, the meter can be used as an enlarging exposure meter. In this connection the relatively large area of the silicon cell is an asset. When the cell is placed on the base board of the enlarger and the image of the negative is projected on to it, the cell receives the image of a relatively large portion of the negative. Its response is related much more to the *average* density of the negative than is that of a small photo-cell. This makes it easier to determine the required exposure.

The circuit is shown in Fig. 55. It may be worth while to use a second resistor with a switch so that either one can be switched into circuit. This would allow the sensitivity of the meter to be altered to suit bright or dim lighting conditions. If the meter is to be used as a reflected light meter the method of construction shown in Fig. 56 is a convenient one. An

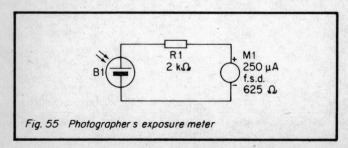

Fig. 55   Photographer s exposure meter

encapsulated solar cell may be mounted on the outer surface of the case but an unprotected cell should be mounted inside a window, with a sheet of perspex or clear polystyrene to protect it. A hood, painted matt black on the inside, restricts the angle of acceptance of the cell, so that it may receive only the light coming from a restricted area of the subject. Fig. 56b shows how the cell and meter may be mounted for an enlarging exposure meter. Again, an unprotected cell should be mounted below a transparent plastic window. There is much to be said for mounting the cell on the *bottom* of the case (whether or not the cell is of the encapsulated type). This brings the sensitive surface close to the plane in which the image is to be focused. A similar layout is

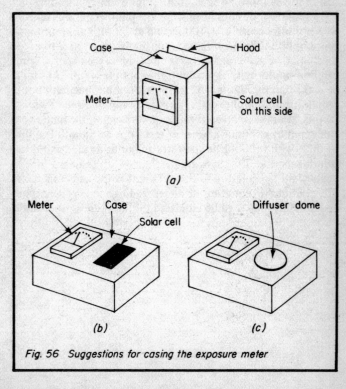

Fig. 56 *Suggestions for casing the exposure meter*

used when the meter is intended for use in the incident light method. In this method the cell is covered by a dome of translucent white material. For a small cell a half of a table-tennis ball may be used. It may be more difficult to find a perfectly hemispherical diffuser for a large cell. In this event a white translucent plastic cap from a aerosol spray may be used instead without serious loss of accuracy. Designs (b) and (c) can of course be combined by making the dome removable, giving a dual-purpose meter.

After the meter has been constructed it needs to be calibrated. This is a matter of taking a number of photographs under various lighting conditions and noting the meter reading, the lens aperture, shutter speed and emulsion speed on each occasion. This will provide sufficient information to allow the dial of the meter to be calibrated. If for example a reading of "3" on the dial was obtained when the photograph was taken with a shutter speed of 1/100 second at f/8 with an emulsion speed of 400ASA, the other combinations appropriate to a reading of 3 can be calculated. For example, the exposure time can be *doubled* to 1/50 second if the aperture is *halved* to f/11 *or* if the film speed is *halved* to 200ASA. An elementary book on photography will explain the relationships between these factors. For use as an enlarging exposure meter, the meter is calibrated by exposing a series of test strips for various lengths of time, with various lens apertures and using a range of printing papers. The strips are developed for a standard time at standard temperature so that different strips are comparable. The relationship between the meter reading and exposure time and aperture may then be tabulated for the various papers used.

CHAPTER 12

The Fun of the Fair

The first two circuits in this chapter provide novelty and light-hearted amusement for the family and could also generate considerable interest and profit at a fund-raising fete.

The Solar Spinner
This device is the electronic equivalent of the "lucky number" spinner. In the original version a pointer is spun around a disc with numbered segments. Players hold tickets or tokens corresponding to each number. Gradually the pointer slows down. Eventually, with mounting excitement among the players, it comes slowly to rest and the winning number is indicated. This solar cell controlled device can be used in the same manner, or as in individual challenge, as will be explained later. Instead of a spinning pointer we have a digital counter that runs through binary count for 0 to 15 (1111) or more in rapid succession. The rate of counting is controlled by a voltage-controlled oscillator (VCO) and the voltage that controls this is derived from a solar cell. Fluctuations in the level of sunlight at the fete (or indoors) add a strong element of chance to the game. The binary numbers are displayed by 4 or more filament lamps (LED's are rather too small to make much impression in this context) and to add to the excitement the output of the VCO is amplified so that it can be heard from a loudspeaker. At high speed a buzzing sound is heard but, as the frequency of the VCO drops, the pitch of the buzz falls lower and lower until individual "clicks" can be distinguished. These come more and more slowly and the lights flash more slowly too (one change of lights for every 8 clicks). Where will the pointer stop? — or what combination of lights will be lit when the oscillator ceases oscillating? The last few clicks come several seconds apart so the suspense is considerable.

The circuit is powered by dry cells or a mains power-pack. As Fig. 57 shows, the oscillator is a CMOS IC, the phase locked loop described in Chapter 10. Here its control

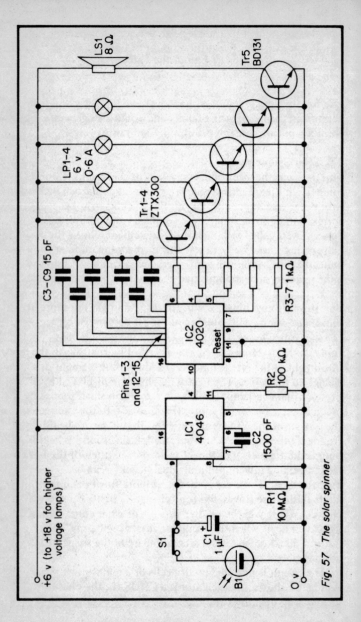

Fig. 57 The solar spinner

94

input is connected to a single solar cell, with a capacitor and resistor in parallel. When the amount of light falling on the cell is reduced or when the push-button S1 is pressed, the charge leaks slowly away from C1 through R1. Voltage falls at an ever decreasing rate, taking about 10—20 seconds to reach zero. The output of the oscillator goes to a counter IC (4020). The first output stage of this IC (pin 9) is at half the frequency of the oscillator and this is fed to Tr5 to produce the clicking sounds in LS1. The next two stages of the dividing chain of the 4020 do not have external terminals. The frequency of the next available stage (pin 7) is 1/8 that of pin 9, which is why there are 8 clicks for every change of the light. The filament lamps are switched on and off by transistors Tr1 — Tr4. The lamps used can be small lamps of the kind found in electric torches or they can be something more decorative, such as Christmas tree lamps. In the latter case a higher supply voltage may be needed. For the fete or fairground the lamps could be mains-powered and operated by relays. A relay-driving circuit is shown in Fig. 58 and four such circuits are required.

There is no need for players to understand about binary numbers in order to play this game. Prizes can be offered for

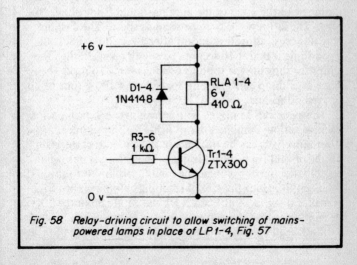

Fig. 58 Relay-driving circuit to allow switching of mains-powered lamps in place of LP 1-4, Fig. 57

certain combinations of lights, as in fruit machines. "All-lamps-on" or "all-lamps-out" are two obvious winning combinations. If the lamps are of several colours, different combinations of colours are awarded prizes with values related to the chances of obtaining each combination. There are many ways of working things out and it is left to the reader to devise the detailed rules.

Begin construction by building the VCO. Its output may be monitored by using an oscilloscope, or connect an earphone to pin 4 by way of an $0.1\mu F$ coupling capacitor. If the rate of fall of frequency is too slow, reduce the value of R1. If the circuit is used with a supply voltage greater than 6v, it will be necessary to decrease the value of C2 or the value of R2 in order that the relatively low voltage produced by B1 produces a sufficiently high frequency. Note that S1 is a *push-to-break* button. Next build the counting section of the circuit, including the switching transistors. To obtain reliable counting, the unused output of IC2 should be coupled to 0v by way of small capacitors, as shown.

When the circuit is switched on, a loud buzz should come from LS1 and the lamps should flash rapidly in a binary sequence. The effect of partially or completely shading B1 should be investigated. When S1 is pressed and held, the rate of flashing and the pitch of the buzzing should fall gradually to zero. The last few clicks come seconds apart. The clicking may not cease altogether except after a very long time, but a good rule is that if 10 seconds pass with no click, then the lamps are showing the winning combination. To make the game even more nerve-racking, extend the waiting time to 20 or even 30 seconds.

The game can be played with several persons, each person backing various combinations of lights for agreed odds. Alternatively, it can be treated as a game requiring a certain degree of skill from the individual. The person states which combination he or she is aiming for and is allowed to shade the solar cell to reduce the rate of counting before pressing S1. It makes thing more difficult if B1 and S1 are mounted 10cm or more apart and the player is allowed to use only *one* hand for both shading *and* button-pressing. Another idea is to place a transparent plastic box over B1, making it impossible for the cell to be completely shaded. Finally, if you prefer a game with really long odds, connect extra lamps to pins 13, 12, 14, 15, 1, 2, 3 (in that order) to give up to 11 lamps with 2^{11} (= 2048) combinations and less than 1 chance in 2000 of obtaining any given one.

Solar Duck-Shoot

This is another game for the home or for the fund-raising fete but, unlike the previous one, it can be run entirely on solar cells. It takes only 250mA at 3v so it can be powered by 6 of the smallest type of cells or 3 cells and the voltage doubler or tripler (p.17). Another way in which it is unlike the previous game is that it requires a greater degree of skill.

The "duck" is the pointer of a microammeter, moving across the dial from left to right. It can be made to move at any speed to suit the level of skill of the contestant. When it reaches the right-hand end of the scale, it flicks quickly back to zero and starts again. The aim of the player is to shoot the duck down when it is exactly at the middle of the scale. A button is provided for shooting and, if the button is pressed at *exactly* the right instant, the duck is shot down. The needle returns instantly to zero instead of continuing across the scale. It is no use to attempt to cheat by pressing the button before the pointer gets to the centre and then holding the button. If the button is pressed even an instant too soon, the needle flashes past the centre and stays well to the right-hand side of the scale until the button is released.

As Fig. 59 shows, this complex-sounding operation can be performed by only a few CMOS ICs. Half of IC1 is used to provide a series of pulses to the clock input of IC2. The action of the pulse generator is described on p.35. IC2 is a 14-stage binary counter. In this circuit we use the outputs of the 4th to 11th stages. These outputs run through the numbers of the binary scale from 00000000 to 11111111 (= 255 decimal) counting at one sixteenth of clock frequency. The output of the first stage (pin 9) is not used; C2 ensures reliability of the counting sequence. The outputs of the 13th and 14th stages are not used either but usually do not need capacitors; if counting is irregular, 15pF capacitors may be connected to these outputs too.

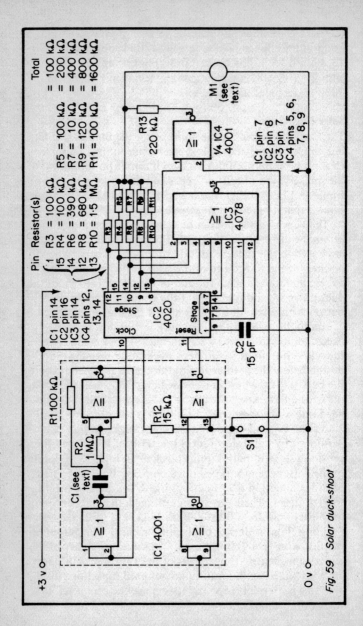

Fig. 59 Solar duck-shoot

As the outputs go through their binary sequence, current flows to the meter from those outputs that are high. Resistors R3 to R11 are chosen so that the total current going to the meter is in proportion to the numerical value of the binary count. The *least* significant digit (stage 8, pin 13) has the *highest* resistance ($1.6M\Omega$) in series with it so that it contributes *least* to the total current. Conversely the *most* significant digit (stage 12, pin 1) has the lowest resistance ($100k\Omega$) in series with it, so that it contributes *most* to the total current. In short, IC2 and resistors R3 to R11 form a digital-to-anologue converter. As counting proceeds, the current to the meter reading increases by small steps. The needle moves gradually across the scale. It returns rapidly to zero when all outputs of IC2 go low at the beginning of the next counting sequence. The needle is half-way along the scale when the count is 128, or 10000000 in binary. This is the stage at which the button S1 should be pressed to shoot the duck. All the inputs to the NOR gate of IC3 are low at that stage. For an instant, until the count becomes 10000001, the output of IC3 is high. At all other times in the cycle (except 00000000, which is not relevant to shooting) it is low. When the output of IC3 is high, the INVERT of this, at pin 10 of IC1, is briefly low. If S1 is pressed at this instant the output at pin 11 goes high, resetting the counter. The return to "all zeroes" shoots down the duck, bringing the pointer immediately back to zero. Note that this logic does not involve the most significant digit, pin 1. It does not matter if this is low or high. This digit is used to detect the pressing of the button *before* the mid-point of the scale is reached. Before the mid-point, the output from pin 1 is low. If the button is pressed then, the output of the NOR gate from IC4 goes high. This output has a low-value resistor in series with it. The needle of the meter is taken well to the right-hand end of the scale, indicating a "miss". If the button is pressed during the second half, when the output of pin 1 is high, there is no effect. Either the duck is "shot" or, if the count has gone on to 10000001 or beyond, the needle proceeds in its normal way to the end of the scale.

In building this circuit construct the pulse generator first. The capacitance chosen for C1 sets the speed at which the duck flies. With a 3v supply a "flight time" of 2 seconds is obtained if C1 has the value $0.022\mu F$. If the supply voltage is 6v, the clock oscillates more rapidly and a capacitor of about $0.047\mu F$ is suitable. The ideal value can be found by experiment: it is easy to modify the circuit to switch in different capacitors, or to use a variable resistor for R1 so that the flying speed may be altered to suit the skill of the player. Next build the digital-to-analogue section and observe the response of the meter. With a $250\mu A$ meter having a 625Ω coil an operating voltage of 6v was found necessary to bring the needle to the far end of the scale at maximum count. This meant that the voltage tripler was required as part of the power supply. For operation on 3v a more sensitive meter should be employed. If there is any problem with the needle over-shooting the end of the scale, a resistor can be wired in series with the meter. At this stage the needle should move up the scale in a series of small but perceptible steps. If steps are irregular test each output stage of IC2 to ensure that it is functioning. Check also that the resistors are of the correct value. Finally wire in the logic circuits, IC3 and IC4 and the two remaining gates of IC1. The game is then ready to play.

Success at this game depends on pressing the button when the count is 10000000, and at no other time. If the pointer takes 3 seconds to traverse the scale, the count stops at 10000000 for only 0.012 seconds. If this is thought to be too difficult a "target" to hit, the speed of flight may be decreased by altering C1. Another course is to reduce the resolution of the logic by ignoring one or more of the least significant digits. If pins 13 or 12 (or both) of IC2 are disconnected from IC3, the unused inputs of IC3 (pins 5, or 4 or both) must be connected to the 0v rail.

One for Luck
This is not a game but a very useful timer that is based on the same circuit as the Solar Duck-Shoot. It is a way of making use of the Duck Shoot when the novelty has worn off. The clock circuit consists of the pulse generator and digital-to-analogue converter but omits the logic circuits (Fig. 60). In this version C1 has a high value so that the counter takes several minutes to go through all its stages. The scale of M1 can

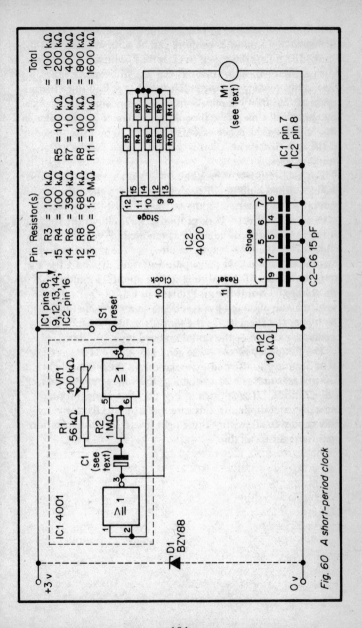

Fig. 60 A short-period clock

101

therefore be graduated to read the time in minutes. A total time-span of 15 minutes or more can be achieved easily. Unlike the timers described in Chapter 7, which indicate only when a given period of time has elapsed, this timer gives a continuous reading of how much time has passed since its run began. It has many applications in the home and office. As a telephone call timer its dial could be graduated not in minutes or seconds, but in pence or tens of pence, to indicate the cost of the call. Different scales would be needed for the various charging rates.

For a description of how the timer works, see the previous project. When building this clock, begin with the pulse generator IC1. Unused inputs to this IC must be connected to 0v or +3v if it is to work properly. VR1 allows the frequency of the generator to be adjusted as required. When choosing value for C1 the total length required for the timing period must be considered. If the supply voltage is 3v, and C1 is a $4.7\mu F$ capacitor, the total time will be about 25 minutes. If the voltage is 6v this time is reduced to half. Other total times may be obtained by selecting a suitable value for C1 — the higher the capacitance the longer the total time. The capacitor must be of the tantalum bead type.

The frequency of the pulse generator depends on voltage, so its accuracy is affected by variations in the level of light. For this reason it is best to stabilise the voltage by connecting a zener diode, D1, as shown in Fig. 60. For 3v operation, use a 3v zener diode. In this case use 7 solar cells to give a 3.5v supply to allow for falling light levels. For 6v operation on 5.6v zener is suitable.

APPENDIX

Data for the Constructor

Diodes

Diagrams of terminal connections of all types mentioned in this book appear in Figure 61. In these diagrams diodes are viewed from the side.

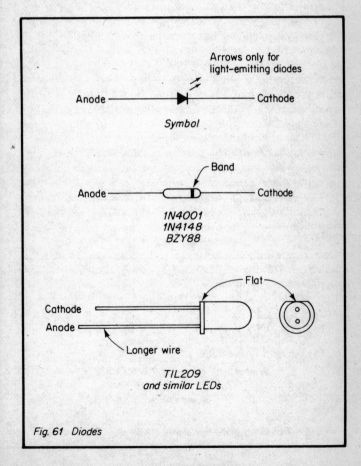

Fig. 61   Diodes

Transistors and Thyristor

Diagrams of terminal connections appear in Fig. 62. In these diagrams the devices are viewed from below.

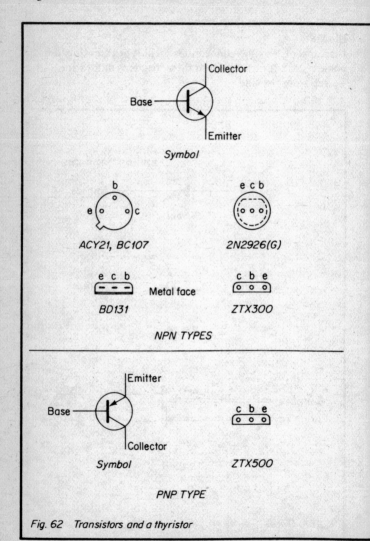

Fig. 62 Transistors and a thyristor

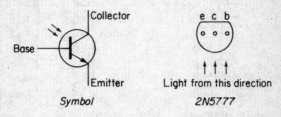

Symbol 2N5777

Light from this direction

PHOTO-DARLINGTON

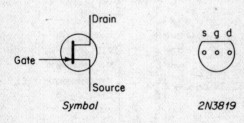

Symbol 2N3819

FIELD-EFFECT TRANSISTOR

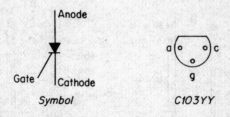

Symbol C103YY

THYRISTOR

Integrated Circuits

In Fig. 63 the CMOS logic ICs are shown as viewed from above.

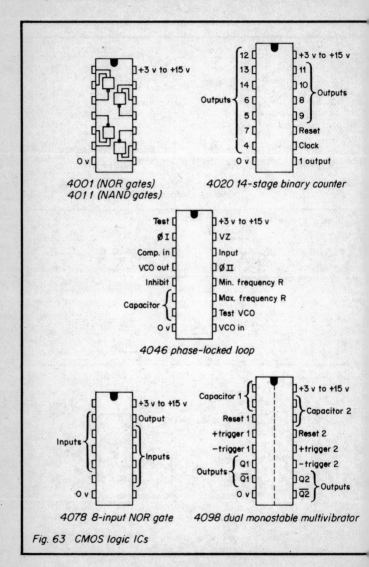

4001 (NOR gates)
4011 (NAND gates)

4020 14-stage binary counter

4046 phase-locked loop

4078 8-input NOR gate

4098 dual monostable multivibrator

Fig. 63 CMOS logic ICs

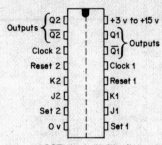

4027 dual JK flip-flop

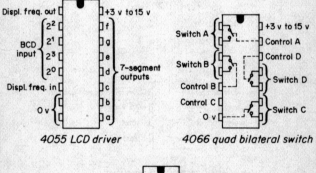

4055 LCD driver

4066 quad bilateral switch

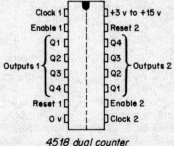

4518 dual counter

In Fig. 64 the DIL integrated circuits are seen from above but the radio IC (ZN414) and the constant current IC (334Z), being similar in appearance to transistors, are viewed from below.

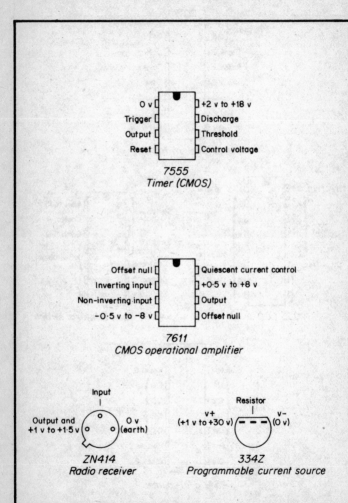

7555
Timer (CMOS)

7611
CMOS operational amplifier

ZN414
Radio receiver

334Z
Programmable current source

Fig. 64 Other integrated circuits

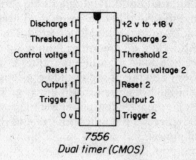

7556
Dual timer (CMOS)

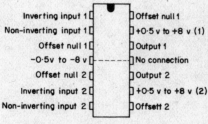

7622
Dual CMOS operational amplifier

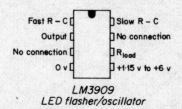

LM3909
LED flasher/oscillator

Suppliers of Components

The components required in this book are available from most mail-order suppliers and component shops. A range of low-priced solar cells is sold by local Tandy stores. The less common components such as the 344Z IC, the low-cost microammeters and touch plates are stocked by Maplin Electronics Ltd., P.O. Box 3, Rayleigh, Essex, SS6 8LR. A very wide range of solar cells of all shapes and sizes is available from Rhienbergs Ltd., Morley Road, Tonbridge, Kent TN9 1RN. They sell everything from ready-assembled arrays of solar cells down to small solar "chips". The latter produce between 1 and 6mA and would be adequate for driving many of the low-power circuits described in this book. The same firm also supplies low-power electric motors (25mA), Fresnel lenses, reflectors and other items that could be of use with solar cell projects.

Notes

Notes

Notes

Notes

Notes

Notes

Please note overleaf is a list of other titles that are available in our range of Radio, Electronics and Computer books.

These should be available from most good Booksellers, Radio Components Dealers and Mail Order Companies.

However, should you experience difficulty in obtaining any title in your area, then please write directly to the publishers enclosing payment to cover the cost of the book plus adequate postage.

If you would like a copy of our latest catalogue of Radio, Electronics and Computer books then please send a Stamped Addressed Envelope to:—

BERNARD BABANI (publishing) LTD
The Grampians
Shepherds Bush Road
London W6 7NF
England